THE OPEN UNIVERSITY

Mathematics: A Third Level Course

Complex Analysis   Units 4, 5 and 6

Integration
Cauchy's Theorem I
Taylor Series

Prepared by the Course Team

The Open University Press

The Open University Press, Walton Hall, Milton Keynes.

First published 1974.

Copyright © 1974 The Open University.

Produced in Great Britain by
Technical Filmsetters Europe Limited, 76 Great Bridgewater Street, Manchester M1 5JY.

ISBN 0 335 05551 6

This text forms part of the correspondence element of an Open University Third Level Course. The complete list of units in the course is given at the end of this text.

For general availability of supporting material referred to in this text, please write to the Director of Marketing. The Open University, P.O. Box 81, Milton Keynes, MK7 6AT.

Further information on Open University courses may be obtained from The Admissions Office, The Open University, P.O. Box 48, Milton Keynes, MK7 6AB.

1.1

CONTENTS

Unit 4 Integration

Conventions

Before working through this text make sure you have read *A Guide to the Course: Complex Analysis.*

References to units of other Open University courses in mathematics take the form:

Unit M100 13, Integration II.

The set Book for the course M231, Analysis, is M. Spivak, *Calculus*, paperback edition (W. A. Benjamin/Addison-Wesley, 1973). This is referred to as:

Spivak.

Optional Material

This course has been designed so that it is possible to make minor changes to the content in the light of experience. You should therefore consult the supplementary material to discover which sections of this text are not part of the course in the current academic year.

4.0 INTRODUCTION

In this unit we shall introduce the complex integral. The theory of analytic functions is completely dependent upon this concept, and we shall therefore spend some time on a careful definition and on a discussion of its basic properties. Since there is a danger that you may lose sight of our main objective among the many minor details that we shall have to discuss, we begin with an intuitive approach.

Intuitive Discussion

A real integral, such as $\int_0^1 \sin x \, dx$, can be regarded as an integral along a "curve", which in this case is the line segment joining 0 and 1. An equally valid integral arising from this function and this line segment is $\int_1^0 \sin x \, dx$. So to determine an integral uniquely we must know if we are to go from 0 to 1, or from 1 to 0.

The complex integral is a generalization of this idea in two ways; firstly, because we replace the real function by a complex function, and secondly, because we replace the straight line segment by a "curve". We shall say exactly what we mean by a curve later, but for the moment let us rely on our intuition. Suppose that we are given a complex function f and a curve Γ. Just as for real integrals, it is essential for us to know the direction that we must go along Γ. We shall indicate the direction with an arrow as in Fig. 1.

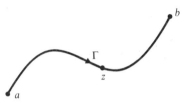

Fig. 1

After a certain amount of effort, towards the end of this unit, we shall be able to give a precise meaning to

$$\int_\Gamma f,$$

but let us first see where our intuition leads us. We can imagine the point z moving along the curve Γ from a to b (in the direction of the arrow) so that at any time t, say, it would be at the point $\gamma(t)$ on the curve: thus $z = \gamma(t)$ at time t. If the point is at a when the time is t_0, and at b when the time is t_1, then $a = \gamma(t_0)$ and $b = \gamma(t_1)$. Notice that γ is a function from $\mathbf{R}$ (or a subset of $\mathbf{R}$) to $\mathbf{C}$, that is, γ is a complex-valued function of a real variable.

If we use Leibniz notation there is an almost irresistible temptation to write

$$\int_\Gamma f = \int_\Gamma f(z) \, dz = \int_{t_0}^{t_1} f(z) \frac{dz}{dt} \, dt = \int_{t_0}^{t_1} f(\gamma(t)) \gamma'(t) \, dt.$$

As it happens the notation is well chosen because the expression on the right is almost our definition of the integral, and it remains only to interpret this expression as the sum of a few real integrals. This we can do quite easily by putting $f(z) = U(x, y) + iV(x, y)$ and $\gamma(t) = \phi(t) + i\psi(t)$, and we then have

$$\int_{t_0}^{t_1} f(\gamma(t)) \gamma'(t) dt = \int_{t_0}^{t_1} [U(\phi(t), \psi(t)) + iV(\phi(t), \psi(t))][\phi'(t) + i\psi'(t)] dt, \quad (*)$$

which can easily be expanded out to give a sum of four real integrals. This will in fact be our starting point in our later definition.

The following example is intended to illustrate our intuitive discussion.

Example

Evaluate $\int_\Gamma f$ where $f(z) = z^2$ and Γ is the semi-circle shown in Fig. 2.

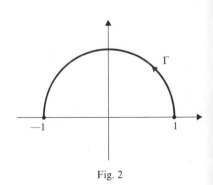

Fig. 2

7

Let $\gamma(t) = \cos t + i \sin t, t \in [0, \pi]$; then

$$\gamma'(t) = -\sin t + i \cos t$$

and

$$f(\gamma(t)) = (\cos t + i \sin t)^2.$$

So

$$\int_\Gamma f = \int_0^\pi f(\gamma(t))\gamma'(t)dt$$

$$= \int_0^\pi (\cos t + i \sin t)^2(-\sin t + i \cos t)dt$$

$$= i \int_0^\pi (\cos t + i \sin t)^3 dt$$

$$= i \int_0^\pi (\cos 3t + i \sin 3t)dt, \quad \text{by de Moivre's Theorem,}$$

$$= -\tfrac{2}{3}.$$

Before going further we should point out two things. The above expression (*) for the integral is nowhere near as bad as it looks, and although it forms the basis for our later definition, we very rarely calculate complex integrals this way. Secondly, you are wasting your time if you try to interpret the integral as an "area under a curve"; its importance is not the physical interpretation that we are able to give it but its use as a device for discovering the properties of analytic functions.

Several problems of a technical nature may well have occurred to you. For example:

(i) Does it matter which function γ we use to determine the position of z at time t? (If it does, then it means that $\int_\Gamma f$ depends not only on the curve Γ but also on how the point z moves along Γ.)

(ii) What exactly do we mean by a "curve" Γ?

(iii) Are we sure what $\gamma'(t)$ means?

(iv) Are we sure that the real functions involved are integrable?

(v) Does $\int_\Gamma f$ depend on the curve Γ, or just on its end-points?

(vi) What conditions on f and Γ ensure that $\int_\Gamma f$ exists?

The purpose of this unit is to answer these questions, and thereby to define *contour integral*. We have set out our programme of study below. We recommend that you read it through *at the start of each reading section* so that you know exactly the stage in the development that we have reached.

The unit can be broken down into several distinct steps.

1. We take our first step towards defining $\int_\Gamma f$ by supposing that f is a complex function, but that Γ is a very simple curve, namely an interval of the real line (Section 4.1).

2. Our next step is to introduce the idea of an arc, which is simply a function γ such that $\gamma(t)$ moves along a "smooth curve" Γ as t varies. We investigate the various different arcs (functions) which correspond to the *same* curve. The point is that different functions γ_1 and γ_2 may take us along the same curve but at different rates or in opposite directions (Section 4.3).

3. We then define the integral of f along a "smooth arc" γ, denoted by $\int_\gamma f$, and examine its properties (Section 4.5).

4. At the next stage we group together all the integrals along arcs which correspond to the same "smooth curve" Γ and so define $\int_\Gamma f$ (Section 4.7).

5. The final step is to piece together "smooth curves" to form *contours*, and then to define the integral along a contour. The *contour integral* is the ultimate goal of the unit, and almost everything which follows in the course depends on it (Section 4.7).

Television and Radio

In the second radio programme associated with the course we shall relate the above programme of study explicitly to the detailed work in the unit.

In the third television programme we shall examine some of the problems encountered in the evaluation of contour integrals.

4.1 FUNCTIONS OF A REAL VARIABLE

This section is concerned with the analysis of functions whose domains are contained in $\mathbf{R}$ and whose ranges are contained in $\mathbf{C}$. In *Unit 2, Continuous Functions*, we called such functions "functions of a real variable"; here we shall often use the description "complex-valued functions of a real variable" for emphasis. We examine functions of a real variable because it is easy to develop a theory of integration for these functions; indeed, the fundamental integration process in complex analysis is the integration of functions of a real variable. So this section has two aims: to explain how to integrate a function of a real variable, certainly; but first, to carry out the groundwork in analysis that we need for the theory of integration.

$$* \qquad * \qquad * \qquad * \qquad * \qquad * \qquad * \qquad *$$

We shall restrict our attention to functions whose domains are closed intervals $[a, b]$ of the real line.

It is easy to invent examples of complex-valued functions of a real variable. Here are some:

$$\phi_1(t) = t(1 + i), \qquad t \in [0, 1];$$

$$\phi_2(t) = t^2(1 + i), \qquad t \in [0, 1];$$

$$\phi_3(t) = e^{it}, \qquad t \in [0, 2\pi].$$

One obvious way of constructing functions of a real variable is to restrict complex functions to the real axis. Thus if f is a complex function whose domain contains the closed interval $\{z : \operatorname{Im} z = 0, \ a \leqslant \operatorname{Re} z \leqslant b\}$, and if ϕ is given by $\varphi(t) = f(t)$ for $t \in [a, b]$ (where t is of course a *real* number), then ϕ is a function of a real variable. All three examples above may be constructed in just this way: for example, ϕ_3 is the restriction of the function $z \longrightarrow e^{iz}$ to the closed interval $[0, 2\pi]$ of the real axis.

When we use phrases such as "real number" and "the real line", we do not usually specify whether we are considering the reals as quite separate from the complex numbers, or as a subset of them. This ambiguity can be very useful. For example, if we think of the real numbers as a subset of the complex numbers we may use the general form of the definitions of limit and continuity, explained in *Unit 2*, as definitions of limit and continuity for functions of a real variable. We write the definition of continuity out below as a reminder.

Definition

> The complex-valued function ϕ of a real variable is **continuous at** $t_0 \in [a, b]$ if for all $\varepsilon > 0$, there is $\delta > 0$ such that for all $t \in [a, b]$ with $|t - t_0| < \delta$, $|\phi(t) - \phi(t_0)| < \varepsilon$.
>
> If ϕ is continuous at all points of $[a, b]$, we say that ϕ is **continuous on** $[a, b]$.

The sum and product of continuous functions are again continuous, as is their quotient if the denominator is non-zero, but, as always, we have to be a little careful when discussing composition to make sure that the composites are defined. If ϕ is a function of a real variable and f is a complex function whose domain contains the range of ϕ, then we can form $f \circ \phi$. It will be a function of a real variable, and will be continuous if both f and ϕ are continuous. Alternatively, if ϕ is a function of a real variable and θ a real function, then $\phi \circ \theta$ is a function of a real variable; it will be continuous if both ϕ and θ are. The function ϕ_2 of our list of examples above is a composite of this type: $\phi_2 = \phi_1 \circ \theta$, where $\theta(t) = t^2$. As an example of the first kind of composition, we observe that since the real and imaginary part functions are continuous, if ϕ is a continuous function of a real variable, then $\operatorname{Re} \phi$ and $\operatorname{Im} \phi$ are both continuous. The converse is also true: if U and V are real functions continuous on the closed interval $[a, b]$, then $t \longrightarrow U(t) + iV(t)$ is a continuous function of a real variable.

We leave the proof to you: see Problem 3 of Section 4.2. (You will recall that to say that a real function ρ is continuous on a closed interval $[a, b]$ means that ρ is continuous on (a, b), and that $\lim_{t \to a^+} \rho(t) = \rho(a)$ and $\lim_{t \to b^-} \rho(t) = \rho(b)$.)

Differentiation

We consider next the differentiation of functions of a real variable.

Definition

The complex-valued function ϕ of a real variable is **differentiable at** $t_0 \in [a, b]$ if $\lim_{h \to 0} \dfrac{\phi(t_0 + h) - \phi(t_0)}{h}$ exists; we call the value of the limit the **derivative of** ϕ **at** t_0 and denote it by $\phi'(t_0)$. To be more explicit, ϕ is differentiable at t_0 and has derivative $\phi'(t_0)$ if for all $\varepsilon > 0$, there is $\delta > 0$ such that whenever $t_0 + h \in [a, b]$ and $0 < |h| < \delta$, then

$$\left| \frac{\phi(t_0 + h) - \phi(t_0)}{h} - \phi'(t_0) \right| < \varepsilon.$$

This definition becomes clearer if we consider its implications for the real and imaginary parts of ϕ.

Theorem 1

The complex-valued function ϕ of a real variable is differentiable on $[a, b]$ if and only if the real functions $\text{Re } \phi = U$ and $\text{Im } \phi = V$ are differentiable on (a, b) and have right-hand derivatives at a and left-hand derivatives at b. Also

$$\phi'(t) = U'(t) + iV'(t).$$

We omit the proof. (If you have time you could construct the proof, which is not difficult.)

The theorem tells us that we may differentiate a function of a real variable by differentiating its real and imaginary parts separately. The next theorem gives another useful method of calculating derivatives. It covers the case in which the function of a real variable is obtained by restriction from an analytic complex function. This is the case we usually have to deal with.

Theorem 2

Suppose that f is a complex function which is analytic on some region containing the interval $[a, b]$ of the real axis, and let ϕ be given by $\phi(t) = f(t)$ for $t \in [a, b]$. Then the complex-valued function ϕ of a real variable is differentiable on $[a, b]$, and $\phi'(t) = f'(t)$ for $t \in [a, b]$.

Proof

Since f is analytic there is $\delta > 0$ such that for all *complex* numbers h for which $0 < |h| < \delta$, we have

$$\left| \frac{f(t + h) - f(t)}{h} - f'(t) \right| < \varepsilon.$$

In particular, if $t + h \in [a, b]$ and $0 < |h| < \delta$, then

$$\left| \frac{\phi(t + h) - \phi(t)}{h} - f'(t) \right| = \left| \frac{f(t + h) - f(t)}{h} - f'(t) \right| < \varepsilon.$$

Thus ϕ is differentiable at t and $\phi'(t) = f'(t)$. ∎

In effect, this theorem says that we may apply the ordinary rules of differentiation to calculate the derivative of a function of a real variable. Thus the derivatives of the functions

$$\phi_1(t) = t(1 + i) \qquad t \in [0, 1],$$

$$\phi_2(t) = t^2(1 + i), \qquad t \in [0, 1],$$

$$\phi_3(t) = e^{it}, \qquad t \in [0, 2\pi],$$

which we used as examples previously, are

$$\phi_1'(t) = (1 + i), \qquad t \in [0, 1],$$

$$\phi_2'(t) = 2t(1 + i), \qquad t \in [0, 1],$$

$$\phi_3'(t) = ie^{it}, \qquad t \in [0, 2\pi].$$

(Note that we use $\phi'(t)$ to denote the derivative of ϕ at t even when t is an endpoint of the interval.)

The usual rules for differentiating sums, products and quotients of functions of a real variable still apply, and we need only take a little care in the statement of the Chain Rule, to allow for the two sorts of composition that can occur when one of the functions, ϕ, is a function of a real variable.

Theorem 3

Let $[a, b]$ be a closed interval.

(i) If $\phi : [a, b] \longrightarrow \mathbb{C}$ is differentiable at t_0 and f is a complex function differentiable at $\phi(t_0)$, then $f \circ \phi$, a complex-valued function of a real variable, is differentiable at t_0 and its derivative is

$$f'(\phi(t_0)) \cdot \phi'(t_0).$$

(ii) If f is a real function whose range contains $[a, b]$ which is differentiable at t_0 and $\phi : [a, b] \longrightarrow \mathbb{C}$ is differentiable at $f(t_0)$, then $\phi \circ f$, a complex-valued function of a real variable, is differentiable at t_0 and its derivative is

$$\phi'(f(t_0)) \cdot f'(t_0).$$

We omit the proof since it is similar to the proof of the Chain Rule for complex functions we gave in *Unit 3, Differentiation*.

Self-Assessment Questions

1. Let f be any function analytic on a region containing the circle $|z| = 1$. Define the function ϕ by $\phi(t) = f(e^{it})$, $t \in [0, 2\pi]$. Show that ϕ is differentiable on $[0, 2\pi]$, and state its derivative. Evaluate $\phi(0)$, $\phi(2\pi)$, $\phi'(0)$ and $\phi'(2\pi)$.

2. Let functions ϕ_1 and θ be defined by

$$\phi_1(t) = t(1 + i), t \in [0, 1], \text{ and } \theta(t) = t^2, t \in [0, 1].$$

Determine $(\phi_1 \circ \theta)(t)$ and use the Chain Rule to compute its derivative.

Solutions

1. ϕ is differentiable on $[0, 2\pi]$ by Theorem 3. Also, by Theorem 3,

$$\phi'(t) = f'(e^{it}) \cdot ie^{it}, \quad t \in [0, 2\pi].$$

So

$$\phi'(0) = ie^{i0} \cdot f'(e^{i0}) = if'(1)$$

and

$$\phi'(2\pi) = ie^{i2\pi} \cdot f'(e^{i2\pi}) = if'(1).$$

Also, since $t \longrightarrow e^{it}$ has period 2π, $\phi(0) = \phi(2\pi) = f(1)$.

2. $(\phi_1 \circ \theta)(t) = t^2(1 + i), t \in [0, 1]$. By the Chain Rule (Theorem 3),

$$(\phi_1 \circ \theta)'(t) = \phi_1'(\theta(t)) \cdot \theta'(t) = (1 + i) \cdot 2t.$$

Integration

All the work we have done so far has been very similar to work we have already done in *Units 2* and *3*. We turn now to integration.

Suppose that ϕ is a complex-valued function of a real variable and $\text{Re } \phi = U$ and $\text{Im } \phi = V$ are both integrable on $[a, b]$, in the sense of real analysis. Then we can define $\int_a^b \phi$ in an obvious way—as $\int_a^b U + i \int_a^b V$, since $\phi = U + iV$.

Definition

> The complex-valued function ϕ of a real-variable is **integrable on** $[a, b]$ if ϕ is defined on $[a, b]$ and its real part U and its imaginary part V are integrable on $[a, b]$. We define its **integral** by
>
> $$\int_a^b \phi = \int_a^b U + i \int_a^b V.$$

If we use Leibniz notation the above equation becomes

$$\int_a^b \phi(t)\, dt = \int_a^b U(t)\, dt + i \int_a^b V(t)\, dt.$$

It follows that the sum of two integrable functions is integrable, and that the "integral of the sum is the sum of the integrals". If ϕ is continuous, then so are U and V; they are therefore integrable; so any continuous function is integrable on any closed interval of its domain.

Self-Assessment Question 3

Calculate the integrals on their domains of the three functions we have used previously.

(i) $\phi_1(t) = t(1 + i)$, $\quad t \in [0, 1]$;

(ii) $\phi_2(t) = t^2(1 + i)$, $\quad t \in [0, 1]$;

(iii) $\phi_3(t) = e^{it}$, $\qquad t \in [0, 2\pi]$.

Solution

(i) If $\phi_1(t) = t(1 + i)$, $t \in [0, 1]$, then

$$\int_0^1 \phi_1 = \int_0^1 t\, dt + i \int_0^1 t\, dt$$

$$= \tfrac{1}{2}t^2 \Big|_0^1 + i \cdot \tfrac{1}{2}t^2 \Big|_0^1$$

$$= \tfrac{1}{2} + \tfrac{1}{2}i.$$

(Note that we have used the notation $\tfrac{1}{2}t^2 \big|_0^1$. In doing this we follow **Spivak** and M231, Analysis.)

(ii) If $\phi_2(t) = t^2(1 + i), t \in [0, 1]$, then

$$\int_0^1 \phi_2 = \int_0^1 t^2 \, dt + i \int_0^1 t^2 \, dt$$

$$= \tfrac{1}{3}t^3 \Big|_0^1 + i \cdot \tfrac{1}{3}t^3 \Big|_0^1$$

$$= \tfrac{1}{3} + \tfrac{1}{3}i.$$

(iii) If $\phi_3(t) = e^{it}, t \in [0, 2\pi]$, then

$$\int_0^{2\pi} \phi_3 = \int_0^{2\pi} \cos t \, dt + i \int_0^{2\pi} \sin t \, dt$$

$$= \sin t \Big|_0^{2\pi} + i \cdot (-\cos t) \Big|_0^{2\pi}$$

$$= 0.$$

Although our definition of the integral is given in terms of real and imaginary parts, it is possible to integrate directly. In fact, as is suggested by Self-Assessment Question 3, the Fundamental Theorem of Calculus has an analogue in complex integration.

Theorem 4

If $g = f'$, where the complex-valued functions g and f of a real variable are defined on $[a, b]$, and f' is integrable on $[a, b]$, then

$$\int_a^b g = f(b) - f(a).$$

Proof

Let $\operatorname{Re} f = U$ and $\operatorname{Im} f = V$. Then $f' = U' + iV'$, and U' and V' are integrable on $[a, b]$. Thus by the Second Fundamental Theorem of Calculus (see **Spivak**, Theorem 14–2),

$$\int_a^b g = \int_a^b U' + i \int_a^b V'$$

$$= U(b) - U(a) + i[V(b) - V(a)]$$

$$= f(b) - f(a). \quad \blacksquare$$

This theorem gives us a technique for evaluating certain real integrals. It sometimes happens that, although it is difficult to integrate some real function, this function is the real part of a complex-valued function which is easy to integrate. We then use the result (which you should establish) that

$$\int_a^b \operatorname{Re} f = \operatorname{Re} \int_a^b f.$$

Example

$\int_0^1 e^{at} \cos bt \, dt$, where a and b are non-zero real numbers, may be evaluated by integration by parts. However, it is simpler to notice that $e^{at} \cos bt$ is the real part of $e^{\alpha t}$ where $\alpha = a + ib$, so that $\int_0^1 e^{at} \cos bt \, dt$ is the real part of $\int_0^1 e^{\alpha t} \, dt$.

Now

$$\int_0^1 e^{\alpha t} \, dt = \int_0^1 \frac{d}{dt}\left(\frac{1}{\alpha} e^{\alpha t}\right) dt = \frac{1}{\alpha} e^{\alpha t} \Big|_0^1 = \frac{1}{\alpha}(e^{\alpha} - 1).$$

The real part of this complex number is easily found, and we obtain

$$\int_0^1 e^{at} \cos bt \, dt = \frac{1}{a^2 + b^2} [e^a(a \cos b + b \sin b) - a].$$

By equating imaginary parts, we also obtain, as a bonus,

$$\int_0^1 e^{at} \sin bt \, dt = \frac{1}{a^2 + b^2}[e^a(a \sin b - b \cos b) + b].$$

One moral to be drawn from Theorem 4 is that, so far as the process of integrating is concerned, we can effectively ignore the fact that we are dealing with complex-valued functions of a real variable and proceed as though all our integrals were real ones. Indeed, we can even use methods like integration by substitution to evaluate such complex integrals. The use of such methods is justified by reducing the integral to two real integrals.

Summary

We have discussed some of the analytical properties of complex-valued functions of a real variable, described how to integrate such functions, and developed some properties of the integral. Probably the most important result is that the Second Fundamental Theorem of (real) Calculus extends to the complex integral. (These integrals correspond to integrating along a curve which happens to be an interval of the real line.)

Self-Assessment Question 4

Let $\phi(t) = \sin \alpha t$, $t \in [-1, 1]$, where $\alpha = a + ib$ is a non-zero complex number.

(i)   Find the real and imaginary parts of ϕ.

(ii) Show that ϕ is continuous on $[-1, 1]$.

(iii) Find the derivative of ϕ.

(iv) Express $\displaystyle\int_0^1 \phi(t)dt$ in terms of real integrals.

(v)   Find $\displaystyle\int_0^1 \phi(t)dt$, by using the Fundamental Theorem, and hence evaluate two real integrals.

Solution

(i) $\sin \alpha t = \sin(at + ibt)$, where $\alpha = a + ib$,

$$= \sin at \cos ibt + \cos at \sin ibt$$

$$= \sin at \cosh bt + i \cos at \sinh bt.$$

Thus, $\operatorname{Re} \phi(t) = U(t) = \sin at \cosh bt$ and $\operatorname{Im} \phi(t) = V(t) = \cos at \sinh bt$.

(ii) Since U and V are continuous on $[-1, 1]$, so is ϕ.

(iii) We may calculate the derivative of ϕ by calculating the derivatives of U and V; but it is much quicker to observe that ϕ is the restriction of the function $z \longrightarrow \sin \alpha z$ to the interval $[-1, 1]$ of the real axis, and then use Theorem 2 to deduce that $\phi'(t) = \alpha \cos \alpha t$.

(iv) $\displaystyle\int_0^1 \phi(t)dt = \int_0^1 \sin at \cosh bt \, dt + i \int_0^1 \cos at \sinh bt \, dt.$

(v) $\displaystyle\int_0^1 \sin \alpha t \, dt = \int_0^1 \frac{d}{dt}\left(-\frac{1}{\alpha}\cos \alpha t\right) dt$

$$= -\frac{1}{\alpha}\cos \alpha t\Big|_0^1$$

$$= \frac{1}{\alpha}(1 - \cos \alpha)$$

$$= \frac{a - ib}{a^2 + b^2}[1 - (\cos a \cosh b - i \sin a \sinh b)]$$

$$= \frac{1}{a^2 + b^2}[a(1 - \cos a \cosh b) + b \sin a \sinh b]$$

$$- \frac{i}{a^2 + b^2}[b(1 - \cos a \cosh b) + a \sin a \sinh b].$$

Equating real and imaginary parts (with those in part (iv)), we obtain

$$\int_0^1 \sin at \cosh bt \, dt = \frac{1}{a^2 + b^2}[a(1 - \cos a \cosh b) + b \sin a \sinh b]$$

and

$$\int_0^1 \cos at \sinh bt \, dt = \frac{-1}{a^2 + b^2}[b(1 - \cos a \cosh b) + a \sin a \sinh b].$$

16

4.2 PROBLEMS

1. Evaluate

$$\int_0^{\pi/2} x \cos 2x \, dx \quad \text{and} \quad \int_0^{\pi/2} x \sin 2x \, dx.$$

2. If f is a complex-valued function of a real variable which is integrable on $[-b, -a]$, show that

$$\int_{-b}^{-a} f(t) \, dt = \int_a^b f(-t) \, dt.$$

3. Show that if U and V are real functions which are continuous on the closed interval $[a, b]$ then $\phi = U + iV$ is a continuous complex-valued function of a real variable.

4. Show that if ϕ is integrable on $[a, b]$ then it is integrable on $[a, t]$ for any t such that $a \leqslant t \leqslant b$. Show that the function $t \to \int_a^t \phi$ is a continuous function of a real variable.

5. Show that if ϕ is integrable on $[a, b]$ and $a < c < b$, then

$$\int_a^c \phi + \int_c^b \phi = \int_a^b \phi.$$

6. Guess, and prove, a rule for integration by parts for complex-valued functions of a real variable.

7. Let ϕ be a continuous function of a real variable defined on the interval $[a, b]$ of the real line, and suppose that $\int_a^b \phi$ is non-zero. Let $\int_a^b \phi = re^{i\theta}$. Show that

$$r = \int_a^b \text{Re}(e^{-i\theta}\phi(t)) \, dt.$$

Use this result to show that

$$\left| \int_a^b \phi \right| \leqslant \int_a^b |\phi|.$$

The result of this problem will be used in Section 4.9, and indeed is the basis for a kind of estimate that will occur in many of the theoretical arguments in the rest of the course.

17

Solutions

1. $\int_0^{\pi/2} x \cos 2x \, dx$ and $\int_0^{\pi/2} x \sin 2x \, dx$ are the real part and imaginary part respectively of

$$I = \int_0^{\pi/2} x e^{2ix} \, dx$$

By integration by parts,

$$I = \left(x \frac{e^{2ix}}{2i} \right)\Big|_0^{\pi/2} - \int_0^{\pi/2} 1 \cdot \frac{e^{2ix}}{2i} \, dx$$

$$= \frac{\pi}{2} \cdot \frac{-1}{2i} - 0 - \left(\frac{e^{2ix}}{-4} \right)\Big|_0^{\pi/2}$$

$$= \frac{i\pi}{4} - \left(\frac{1}{4} - \left(-\frac{1}{4} \right) \right)$$

$$= -\frac{1}{2} + \frac{i\pi}{4}.$$

So $\int_0^{\pi/2} x \cos 2x \, dx = -\frac{1}{2}$ and $\int_0^{\pi/2} x \sin 2x \, dx = \frac{\pi}{4}$.

2. Let $\operatorname{Re} f = U$ and $\operatorname{Im} f = V$. Then

$$\int_{-b}^{-a} f(x) \, dx = \int_{-b}^{-a} U(x) dx + i \int_{-b}^{-a} V(x) dx.$$

Let $x = -t, dx = -dt$, then

$$\int_{-b}^{-a} f(x) dx = -\int_b^a U(-t) dt - i \int_b^a V(-t) dt$$

$$= \int_a^b U(-t) dt + i \int_a^b V(-t) dt$$

$$= \int_a^b f(-t) dt.$$

3. Since U and V are continuous on $[a, b]$, given $\varepsilon > 0$, there are $\delta_1, \delta_2 > 0$ such that

   (i) if $|t - t_0| < \delta_1$, then $|U(t) - U(t_0)| < \dfrac{\varepsilon}{2}$,

   and (ii) if $|t - t_0| < \delta_2$, then $|V(t) - V(t_0)| < \dfrac{\varepsilon}{2}$.

   It follows that if $\delta = \min(\delta_1, \delta_2)$, and if $|t - t_0| < \delta$, then

$$|\phi(t) - \phi(t_0)| = |[U(t) - U(t_0)] + i[V(t) - V(t_0)]|$$

$$\leqslant |U(t) - U(t_0)| + |V(t) - V(t_0)|$$

$$< \frac{\varepsilon}{2} + \frac{\varepsilon}{2} = \varepsilon.$$

   Hence ϕ is continuous.

4. Let $\phi = U + iV$. Then U and V are integrable on $[a, b]$, and therefore also on $[a, t]$. Hence $\phi = U + iV$ is integrable on $[a, t]$, and

$$\int_a^t \phi = \int_a^t U + i \int_a^t V.$$

   Since $t \longrightarrow \int_a^t U$ and $t \longrightarrow \int_a^t V$ are continuous functions, we immediately deduce from Problem 3 that $t \longrightarrow \int_a^t \phi$ is also continuous.

5. Split ϕ into real and imaginary parts, and apply the corresponding result for real integrals to each part.

6. **Integration by parts for complex-valued functions of a real variable.**

   If f and g are differentiable complex-valued functions on $[a, b]$, and if f' and g' are continuous on $[a, b]$, then

$$\int_a^b f g' = f(b)g(b) - f(a)g(a) - \int_a^b g f'.$$

   Proof : Apply Theorem 4 to fg and $(fg)' = fg' + gf'$.

7. If $re^{i\theta} = \int_a^b \phi = \int_a^b \phi(t)dt$, then

$$r = \int_a^b e^{-i\theta}\phi(t)dt, \quad \text{since } e^{-i\theta} \text{ is a constant,}$$

$$= \int_a^b \text{Re}(e^{-i\theta}\phi(t))dt + i\int_a^b \text{Im}(e^{-i\theta}\phi(t))dt.$$

But r is real, so that the second integral on the right hand side must vanish; hence

$$r = \int_a^b \text{Re}(e^{-i\theta}\phi(t))dt, \quad \text{as required}$$

Now,

$$\left| \int_a^b \phi \right| = |re^{i\theta}|$$

$$= r$$

$$= \int_a^b \text{Re}(e^{-i\theta}\phi(t))dt, \quad \text{from above,}$$

$$\leqslant \int_a^b |e^{-i\theta}\phi(t)|dt, \quad \text{since Re } z \leqslant |z| \text{ for all } z,$$

$$= \int_a^b |\phi(t)|dt.$$

4.3 ARCS AND PATHS

In Section 4.1 we discussed functions of a real variable. We turn now to a geometrical interpretation of such functions. That there is a ready geometrical interpretation becomes clear as soon as we look at the ranges of some complex-valued functions of a real variable. Let us use the same examples that we had at the beginning of Section 4.1.

Examples

1. The range of $\phi_1(t) = t(1 + i)$, $t \in [0, 1]$, is the line segment $[0, 1 + i]$ joining the origin to the point $1 + i$.

   If we regard t as representing time then we can say that as t increases from 0 to 1, the point $\phi_1(t)$ moves steadily from 0 to $1 + i$ (Fig. 3).

2. The range of $\phi_2(t) = t^2(1 + i)$, $t \in [0, 1]$, is also the segment $[0, 1 + i]$.

   As t increases from 0 to 1, the point $\phi_2(t)$ moves from 0 to $1 + i$, at a different rate from that of $\phi_1(t)$ (Fig. 4).

3. The range of $\phi_3(t) = e^{it}$, $t \in [0, 2\pi]$, is the circle $\{z:|z| = 1\}$.

   As t increases from 0 to 2π, the point $\phi_3(t)$ moves once round the circle in the direction shown in Fig. 5, starting and ending at 1.

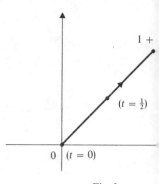

Fig. 3

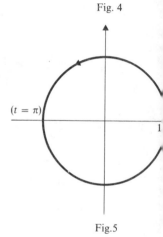

Fig. 4

Note that there is a distinction to be drawn between a curve considered as a set of points in the plane, and a curve considered as the trace of a moving point. A complex-valued function of a real variable gives the latter. In our definitions we are careful to preserve this distinction, and we do so by using the following terminology.

Definition

> An **arc** is a continuous function γ of a real variable with domain a closed interval $[a, b]$ of the real line. The set of points $\{\gamma(t):t \in [a, b]\}$ is called the **path of the arc** γ. The point $\gamma(a)$ is called the **initial point**, and $\gamma(b)$ the **final point**, of the path. We also call an arc with given path a **parametrization** of the path

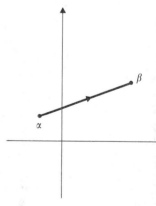

Fig.5

The word *curve* will be used only when it is not important to make the distinction between arc and path.

These definitions are, in part, just modified versions of definitions to be found in *Unit M231 5, Differentiation*.

Here are two important kinds of arc.

Line Segment

If

$$\gamma_1(t) = \frac{b - t}{b - a} \cdot \alpha + \frac{t - a}{b - a} \cdot \beta,$$

Fig. 6

where $t \in [a, b]$, and α and β are complex numbers, $\beta \neq \alpha$, then the path of γ_1 is the line segment $[\alpha, \beta]$ (Fig. 6). The initial point is $\gamma_1(a) = \alpha$ and the final point is $\gamma_1(b) = \beta$. So γ_1 is a parametrization of $[\alpha, \beta]$.

Circle

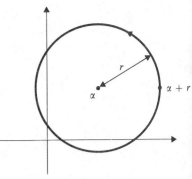

Consider next the arc $\gamma_2(t) = \alpha + re^{it}$, where $t \in [0, 2\pi]$, α is a complex number, and r is a positive real number. The path of γ_2 is the circle centre α, radius r. The initial and final points coincide at $\alpha + r$. We could also have specified this path in set notation, namely $\{z : |z - \alpha| = r\}$.

One–one arcs are particularly important. The path of such an arc can have no crossings or self-intersections.

Fig. 7

Definition

An arc which is one–one is said to be **simple**.

In other words, an arc γ is simple if $\gamma(t_1) = \gamma(t_2)$ only when $t_1 = t_2$.

The arc $\gamma_2(t) = \alpha + re^{it}, t \in [0, 2\pi]$, is not simple because $\gamma_2(0) = \gamma_2(2\pi)$. We call an arc such as this, for which initial and final points coincide, a closed arc.*

Definition

An arc γ with domain $[a, b]$ for which $\gamma(a) = \gamma(b)$ is said to be **closed**.

A closed arc cannot be simple. But the arc γ_2 is as nearly simple as a closed arc can be; it is one–one on $[0, 2\pi)$. We have a special, if slightly contradictory, terminology for such arcs.

Definition

A closed arc γ with domain $[a, b]$ such that γ is one–one on $[a, b)$, is said to be a **simple-closed arc**.

The following diagram illustrates the *paths* of these various kinds of arc.

(i) simple   (ii) not simple, not closed   (iii) simple–closed   (iv) closed, but not simple–closed

Fig. 8

One word of warning: the properties we have defined are properties of arcs, *not* paths. Thus Fig. 8(i) shows a curve which is the path of a simple arc; but it is also the path of many arcs which are not simple. Similarly, Fig. 8(iii) shows the path of a simple-closed arc, say $t \longrightarrow e^{it}, t \in [0, 2\pi]$; but this curve is also the path of $t \longrightarrow e^{it}, t \in [0, 3\pi]$, which is neither simple nor closed. (Think of the motion of a point traversing the circle according to this parametrization.) However, we may safely say that the curve in Fig. 8(ii) cannot be the path of a simple arc, and Fig. 8(iv) cannot be the path of a simple-closed arc.

Another important property of arcs is concerned with differentiability.

Definition

A non-closed arc is **smooth** if it is differentiable on its domain $[a, b]$ and its derivative is continuous and not zero on $[a, b]$.

---

* This is yet another meaning for the word "closed". The context will always make it clear which meaning is intended.

21

It is interesting to consider the following arc in the light of this definition. Let

$$\gamma(t) = \begin{cases} t^2, & -1 \leqslant t \leqslant 0 \\ it^2, & 0 \leqslant t \leqslant 1. \end{cases}$$

Then γ is differentiable everywhere on $[-1, 1]$, and its derivative is continuous; but $\gamma'(0) = 0$, so γ is not a smooth arc. This is just as well, as we see when we examine the path of γ (Fig. 9).

In fact, if we think of t as time, and $|\gamma'(t)|$ as the speed of a point at time t, then the speed of the point at the corner is clearly zero; in other words, the point slows right down as it reaches the corner, and then accelerates out of it.

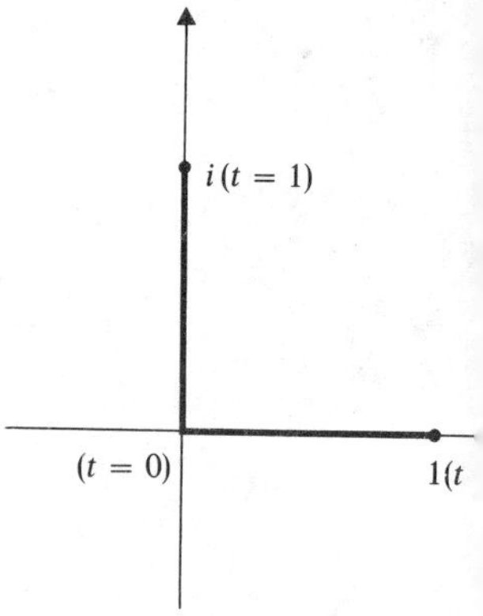

As it stands, our definition of "smooth" does not allow us to talk of a smooth *closed* arc. The reason is that we impose an extra condition on a closed arc γ; we require that the derivative at the initial and final points of γ must be equal for γ to be smooth. So a closed arc γ is **smooth** if it is differentiable on its domain $[a, b]$, its derivative is continuous and not zero on $[a, b]$ and $\gamma'(a) = \gamma'(b)$. The circle $\gamma(t) = e^{it}$, $t \in [0, 2\pi]$, is a simple-closed smooth arc. The path of a closed arc which is not smooth is shown in Fig. 10.

Path of γ

Fig. 9

Fig. 10

Reparametrization

For many purposes the specification of an arc gives too much information, whereas that of a path gives too little. We now aim for a notion intermediate between arc and path and begin by examining the question: "Which arcs have the same path?" or, equivalently: "What different parametrizations are there of a given path?"

Suppose that γ is an arc defined on an interval $[c, d]$ and that θ is a continuous real function from $[a, b]$ onto $[c, d]$. Then $\gamma \circ \theta$ is an arc which has the same path as γ. For example, if $\theta : [0, 1] \longrightarrow [0, 1]$ is given by $\theta(t) = t^2$, and $\gamma(t) = t(1 + i)$ for $t \in [0, 1]$, then $(\gamma \circ \theta)(t) = t^2(1 + i)$.

Definition

> If γ is an arc with domain $[c, d]$ and θ is a continuous real function from $[a, b]$ onto $[c, d]$, we call the arc $\gamma \circ \theta$ the **reparametrization of the path of γ by θ**.

It is natural to ask what conditions on a reparametrization ensure that properties such as being simple or being smooth are preserved.

If γ is a simple arc with domain $[c, d]$ and $\theta : [a, b] \longrightarrow [c, d]$ is continuous and one–one, then $\gamma \circ \theta$ is simple. (Remember that a continuous one–one real function defined on a closed interval is either increasing or decreasing.) If θ is increasing, then $\theta(a) = c$ and $\theta(b) = d$; in this case γ and $\gamma \circ \theta$ have the same initial point and the same final point. In fact they represent points traversing the same curve with different speeds, but in the same direction. On the other hand, if θ is decreasing then $\gamma \circ \theta$ represents a point traversing the path of γ in the opposite direction.

If γ is a smooth arc with domain $[c, d]$ and $\theta : [a, b] \longrightarrow [c, d]$ is continuously differentiable and its derivative is not zero, then $\gamma \circ \theta$ is also smooth. (Note that a real function is said to be *continuously differentiable* if it has a continuous derivative at each point of its domain.)

Reparametrization by an increasing, continuously differentiable function whose derivative does not vanish are important, so we have a special name for them.

Definition

> Let γ be a smooth arc with domain $[c, d]$ and let θ be an increasing, continuously differentiable real function from $[a, b]$ onto $[c, d]$ whose derivative does not vanish. Then we call $\gamma \circ \theta$ a **proper reparametrization of the path of γ by** θ.

The point of proper reparametrizations is that they preserve the "direction" defined by an arc.

Sometimes, however, we shall need an arc which traverses a given path in the opposite direction and so we introduce the following idea.

Let $\gamma : [a, b] \longrightarrow \mathbf{C}$ be a smooth arc. Define $\theta : [a, b] \longrightarrow [a, b]$ by $\theta(t) = (a + b) - t$. Then $\theta'(t) = -1$; and so θ is decreasing. The smooth arc $\gamma \circ \theta$ is a reparametrization of the path of γ, but not of course a proper one, since θ is not increasing. In effect, it defines the direction opposite to that defined by γ. We call $\gamma \circ \theta$ the **reverse** of the arc γ, and denote it by $\tilde{\gamma}$. The reverse of an arc reverses the order of the endpoints, of course:

$$\tilde{\gamma}(a) = \gamma(\theta(a)) = \gamma(b); \quad \tilde{\gamma}(b) = \gamma(\theta(b)) = \gamma(a).$$

The following result will be useful later.

Result

If γ_1 and γ_2 are simple smooth arcs with the same path then it can be shown that γ_2 is a proper reparametrization of either γ_1 or its reverse $\tilde{\gamma}_1$.

We omit the proof of this result which is a little tedious, but not difficult. The same result is true for simple-closed smooth arcs, but in this case we can no longer use the order of the end-points to distinguish one direction from the other. However, for the circle (which is the path of a simple-closed smooth arc) there is no problem: one direction is clockwise and one anticlockwise. We shall agree to call the anticlockwise direction *positive*. (See *Unit M231 11, Techniques of Integration*, Section 11.7.)

In the general case we shall need to be more careful. We could say that we are traversing the path in the positive direction if its inside is on our left. But this apparently innocuous idea raises an extremely tricky point. What is the inside of the path of a simple-closed arc? It is easy to decide in particular cases, but the general result is not at all easy. It is enshrined in the Jordan Curve Theorem, which we shall now state, but regard as too difficult to prove in full generality in this course. See Problem 5 of Section 4.4. (It is so often the case that the simplest things in mathematics turn out to be the most difficult.)

Theorem 5 (The Jordan Curve Theorem)*

If γ is a simple-closed arc, then the set of points in the plane which do not lie on the path of γ is the union of two disjoint regions. One of these regions is bounded: this we call the *inside* of γ. The other, the *outside*, is unbounded. Each point of the path of γ is a boundary point of either region.

Directed Paths

Towards the end of this unit we shall examine integrals along a path with a given direction, and in fact directed paths will play a very important role in the rest of the course.

We can represent directed paths diagrammatically; the addition of a suitable number of arrows to the picture of a path will define a directed path.

---

* *Camille Jordan* (1838-1922) was professor at the École Polytechnique in Paris. He is well known for his work in matrices (the Jordan normal form—see M201), for the Jordan-Hölder Theorem in group theory and for his work in analysis and topology. He published *Cours d'Analyse*, a treatise which was widely read during his lifetime. His own proof of the Jordan Curve Theorem was, in fact, incorrect, and the theorem was not correctly proved until 1905 (by O. Veblen, an American topologist).

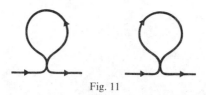

Fig. 11

It will sometimes be necessary for you to choose a parametrization of a directed path specified by a diagram: you must be sure to choose one consistent with the direction of the path.

Examples

4.

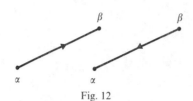

Fig. 12

These are the two possible directions on a line segment. A parametrization consistent with the first would be

$$\gamma(t) = \alpha + t(\beta - \alpha), \qquad t \in [0, 1];$$

one consistent with the second would be

$$\gamma(t) = \beta + t(\alpha - \beta), \qquad t \in [0, 1].$$

5. Here are the two directed paths of a simple-closed smooth arc.

Fig. 13

6. When talking about circles in particular, we assume, unless the contrary is stated explicitly, that we have chosen the positive direction. Thus "a parametrization of the circle $|z| = 1$" is understood to mean "a parametrization of the circle $|z| = 1$ consistent with the positive direction". The standard such parametrization is of course

$$\gamma(t) = e^{it}, \qquad t \in [0, 2\pi].$$

7. Here is a path of an arc which is not simple.

Fig. 14

It has several directions:

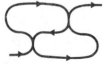 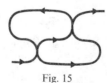 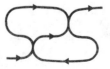

Fig. 15

and of course the reverses of these. (You would not be expected to find a parametrization of this path.)

Hereafter, we shall try to ensure that our figures are sufficiently specific to enable you to determine which direction is intended.

Arc Length

Finally in this section, we wish to remind you of a result from real analysis concerning arc length. If a differentiable arc (U, V) is such that U' and V' are bounded on $[a, b]$ then its length is $\int_a^b \sqrt{(U')^2 + (V')^2}$. (See *Unit M2319, Properties of the Integral*, where arc length and the derivation of this formula are discussed.)

This formula becomes more plausible if we interpret t as time, for then U', V' are the components of instantaneous velocity, and so $\sqrt{(U')^2 + (V')^2}$ is the instantaneous speed of a point describing the path; the formula for arc length is then the appropriate generalization of the rule: distance = speed $\times$ time.

If $\gamma : [a, b] \longrightarrow \mathbf{C}$ is a smooth arc which is a complex-valued function of a real variable, and $\gamma = U + iV$, then $\sqrt{(U')^2 + (V')^2} = |\gamma'|$. Thus we define the **length of a smooth arc** γ to be $\int_a^b |\gamma'|$.

In Problem 4 of Section 4.4 you will be asked to show that arc length is un-affected by a proper reparametrization.

Summary

The things you will most need to remember from this section are: the definitions of, and the distinction between arc and path; the different kinds of arc (simple, simple-closed, smooth); the meaning of reparametrization, directed path, and arc length.

Self-Assessment Questions

1. Let $\gamma_1(t) = t(1 + i), t \in [0, 1]$; let $\gamma_2(t) = t^2(1 + i), t \in [0, 1]$.

    (i) Is γ_1 a smooth arc? Is it a simple arc?

    (ii) Is γ_2 a smooth arc? Is it a simple arc?

    (iii) What are the paths of γ_1 and γ_2?

    (iv) Express γ_2 as a reparametrization of the path of γ_1. Is your reparametrization proper? Can there be a proper reparametrization of the path of γ_1 which gives γ_2?

2. Let $\gamma_n(t) = e^{int}, t \in [0, 2\pi]$, n an integer.

    (i) Show that γ_n is a smooth closed arc for every $n \neq 0$.

    (ii) For which values of n is γ_n simple?

    (iii) For which values of n is γ_n simple-closed?

    (iv) For which values of n does γ_n traverse the positive direction on its path?

    (v) Describe the motion of a point moving along the path of γ_n according to the parametrization γ_n, indicating the differences that occur for different n.

3. Sketch the paths of the arcs γ specified below.

    (i) $\gamma(t) = a \cos t + ib \sin t, \quad t \in [0, 2\pi]$.

    (ii) $\gamma(t) = t + it^2, \quad t \in [-a, a]$.

    Describe these paths geometrically.

4. How many different directions can you find on this path of a closed smooth arc?

Fig. 16

Solutions

1. (i) γ_1 is smooth and simple.

   (ii) γ_2 is not smooth since $\gamma_2'(0) = 0$; it is simple.

   (iii) The path of γ_1 is the line segment $[0, 1 + i]$.
   The path of γ_2 is the line segment $[0, 1 + i]$.

   (iv) $\gamma_2 = \gamma_1 \circ \theta$ where $\theta:[0, 1] \longrightarrow [0, 1]$ and $\theta(t) = t^2$. Since $\theta'(0) = 0$, the reparametrization γ_2 is *not* proper.

   Let γ_2 be a reparametrization of γ_1 by ψ: $\gamma_2 = \gamma_1 \circ \psi$. Hence $\gamma_2(t) = (1 + i)t^2 = (1 + i)\psi(t)$, that is $\psi(t) = t^2$. In other words, there is no proper reparametrization.

2. (i) Since 2π is a period of γ_n, $\gamma_n(0) = \gamma_n(2\pi)$; hence γ_n is closed for all n. Since $\gamma_n'(t) = ine^{int}$, γ_n' is continuous, and γ_n' is non-zero if $n \neq 0$. Also $\gamma_n'(0) = \gamma_n'(2\pi)$. In other words, γ_n is a smooth closed arc for every $n \neq 0$.

   (ii) None.

   (iii) $n = -1, n = 1$.

   (iv) Positive n.

   (v) For $n = 0$, the path is a single point, 1.

   For $n > 0$, the point moves round the circle $|z| = 1$, n times in the positive direction with constant speed $|\gamma_n'(t)| = n$.

   For $n < 0$, the point moves round the circle $|z| = 1$, n times in the negative direction with constant speed $|\gamma_n'(t)| = n$.

3. (i)   (ii)

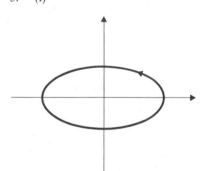

Ellipse $\left\{(x, y): \dfrac{x^2}{a^2} + \dfrac{y^2}{b^2} = 1\right\}$

Fig. 17

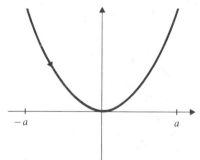

Parabola $\{(x, y): y = x^2, -a < x < a\}$

Fig. 18

4. Four.

26

4.4 PROBLEMS

1. Show that the path of the arc γ defined by

$$\gamma(t) = \begin{cases} -t, & t \in [-1, 0) \\ 0, & t = 0 \\ t + it^3 \sin\left(\dfrac{\pi}{t}\right), & t \in (0, 1] \end{cases}$$

   intersects itself infinitely many times. Is γ a closed arc? Is γ a smooth arc?

2. Consider the arcs γ_1 and γ_2 defined by

$$\gamma_1(t) = (1 - 2t^2) - it, \qquad t \in [-1, 1]$$

   and

$$\gamma_2(t) = \cos 2t - i \sin t, \qquad t \in [-\pi/2, \pi/2].$$

   Show that γ_2 is a reparametrization of the path of γ_1.

3. Use the definition of Section 4.3 to find the lengths of

   (i) the line segment $[0, 1 + i]$;

   (ii) the line segment $[a + ib, x + iy]$;

   (iii) the semi-circular path of the arc γ, where

$$\gamma(t) = ae^{it}, \quad t \in [0, \pi].$$

4. Show that if γ_2 is a proper reparametrization of the path of an arc γ_1 then the arc length is not changed.

5. Prove the Jordan Curve Theorem for a circle and an ellipse.

Problems 6 and 7, whose results will be used later in the course, refer to sets which are path-connected. We define a set S of complex numbers to be **path-connected** if for any points α and β of S, there is an arc γ with domain $[a, b]$ which has α and β as its initial and final points (so that $\gamma(a) = \alpha$, $\gamma(b) = \beta$) and is such that the path of $\gamma \subset S$. Note that since every polygonal line is a path, it follows, by Section 2.7 of *Unit 2*, that every connected set is path-connected; the following converse result is also true: every path-connected open set is connected.

---

6. Prove that every path-connected open set is connected.

7. Show (by an explicit construction) that any annulus $A = \{z : r < |z - a| < R\}$, $r > 0$, is path-connected.

27

Solutions

1. The path intersects itself at the points $t = \dfrac{1}{n}$, $n = 1, 2, 3, \ldots$.

Clearly γ is closed, since $\gamma(-1) = \gamma(1)$. It is not smooth since $\gamma'(t)$ does not exist when $t = 0$.

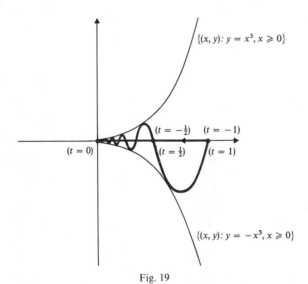

Fig. 19

2. We must find θ such that $\gamma_2 = \gamma_1 \circ \theta$ where $\theta:[-\pi/2, \pi/2] \longrightarrow [-1, 1]$ is continuous. The domain and range of θ suggest $\theta = \sin$, which is continuous. Now

$$(\gamma_1 \circ \theta)(t) = (\gamma_1 \circ \sin)(t)$$
$$= 1 - 2\sin^2 t - i\sin t$$
$$= \cos 2t - i\sin t$$
$$= \gamma_2(t), \quad \text{as required.}$$

3. (i) Using the parametrization $\gamma(t) = t(1 + i)$, $t \in [0, 1]$, we obtain

$$\text{length of } [0, 1 + i] = \int_0^1 |\gamma'(t)|dt$$
$$= \int_0^1 |1 + i|dt$$
$$= \sqrt{2}.$$

(ii) Using the parametrization

$$\gamma(t) = (a + ib) + t[(x + iy) - (a + ib)], \qquad t \in [0, 1],$$

we obtain

$$\text{length of } [a + ib, x + iy] = \int_0^1 |\gamma'(t)|dt$$
$$= \int_0^1 |(x + iy) - (a + ib)|dt$$
$$= [(x - a)^2 + (y - b)^2]^{1/2}.$$

(iii) Using the given parametrization, we obtain

$$\text{length of semi-circle} = \int_0^\pi |\gamma'(t)|dt$$
$$= \int_0^\pi |aie^{it}|dt$$
$$= \pi a.$$

28

4. Let $\gamma_2 = \gamma_1 \circ \theta$ be a proper reparametrization of the path of γ_1 by θ. Let $\gamma_1(t)$ have domain $[a, b]$, and let $\theta:[c, d] \longrightarrow [a, b]$. We wish to show that

$$\int_a^b |\gamma_1'(x)|dx = \int_c^d |\gamma_2'(t)|dt.$$

By the Chain Rule

$$\gamma_2'(t) = \gamma_1'(\theta(t)) \cdot \theta'(t).$$

Hence

$$\int_c^d |\gamma_2'(t)|dt = \int_c^d |\gamma_1'(\theta(t)) \cdot \theta'(t)|dt$$

$$= \int_c^d |\gamma_1'(\theta(t))||\theta'(t)|dt, \quad \text{since } \theta' \text{ is real and positive,}$$

$$= \int_a^b |\gamma_1'(x)|dx, \quad \text{using the substitution } x = \theta(t), \, dx = \theta'(t)dt.$$

5. Since a circle is a special type of ellipse, we shall prove the Jordan Curve Theorem for the latter. Without loss of generality, we can suppose that the centre of the ellipse lies at the origin, and that the ellipse is $\left\{(x, y): \dfrac{x^2}{a^2} + \dfrac{y^2}{b^2} = 1\right\}$: this is the path of the simple-closed arc γ given by $\gamma(t) = a \cos t + b \sin t$, $t \in [0, 2\pi]$. (See Self-Assessment Question 3(i) of Section 4.3). It remains only to identify the two regions into which the path of γ divides the plane; these are clearly $\left\{(x, y): \dfrac{x^2}{a^2} + \dfrac{y^2}{b^2} < 1\right\}$, which is a bounded region—the inside of γ—and $\left\{(x, y): \dfrac{x^2}{a^2} + \dfrac{y^2}{b^2} > 1\right\}$, which is an unbounded region— the outside of γ. These regions are disjoint and their common boundary is the path of γ.

6. Let S be a path-connected open set, and let $S = G_1 \cup G_2$ where G_1 and G_2 are non-empty disjoint open sets. We shall derive a contradiction, thereby proving that S is connected. Let $\alpha \in G_1$ and $\beta \in G_2$, and let γ be any arc with initial point α and final point β such that the path of γ lies in S.

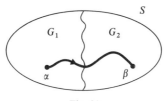

Fig. 20

Consider the function $f : S \to \mathbf{R}$ defined by

$$f(z) = \begin{cases} 0, & \text{if } z \in G_1 \\ 1, & \text{if } z \in G_2. \end{cases}$$

Then, since G_1 and G_2 are open, it is easy to see that f is continuous, and hence so is $f \circ \gamma:[0, 1] \to \mathbf{R}$. But $(f \circ \gamma)(0) = 0$ and $(f \circ \gamma)(1) = 1$, and so, by the Intermediate Value Theorem, $(f \circ \gamma)(t) = \frac{1}{2}$ for some $t \in (0, 1)$, which contradicts the definition of f.

7. It is sufficient to deal with the case $a = 0$.
Let $\alpha = r_1 \exp(i\theta_1)$ and $\beta = r_2 \exp(i\theta_2)$, where r_1 and r_2 both lie between r and R.

We must find an arc lying entirely within the annulus which has α and β as initial and final points (Fig. 21). There are many possibilities—one such is given by

$$\gamma(t) = [r_1 + t(r_2 - r_1)] \exp(i[\theta_1 + t(\theta_2 - \theta_1)]), \qquad t \in [0, 1].$$

Notice that as t increases steadily from 0 to 1, the radius changes steadily from r_1 to r_2 and the argument changes steadily from θ_1 to θ_2.

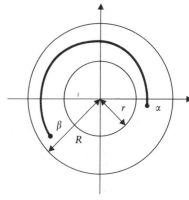

Fig. 21

29

4.5 INTEGRATION ALONG SMOOTH ARCS

Let f be a complex-valued function of a real variable with domain containing the interval $[a, b]$ of the real line. In Section 4.1 we defined the integral of f on $[a, b]$. Now suppose that we think of $[a, b]$ not just as an interval, but as the path of some smooth arc γ (for example, we might take

$$\gamma(t) = a(1 - t) + bt, \quad t \in [0, 1]).$$

Let $[c, d]$ be the domain of γ. Then, by an application of integration by substitution for complex integrals,

$$\int_a^b f(t)dt = \int_c^d f(\gamma(u))\gamma'(u)du.$$

This unremarkable observation gives us a clue how to define the integral of f along *any* smooth arc γ.

Definition

Let γ be a smooth arc with domain $[a, b]$, and f a function continuous on the path of γ. Then the **integral of f along** γ, written as $\int_\gamma f$ or $\int_\gamma f(z)dz$, is defined by

$$\int_\gamma f(z)dz = \int_a^b f(\gamma(t))\gamma'(t)dt.$$

Note that, since f is continuous and γ is smooth, both $f \circ \gamma$ and γ' are continuous on $[a, b]$, and so the integral exists.

Example 1

Evaluate $\int_\gamma z^2 dz$ where

(i) $\gamma(t) = t, \quad t \in [a, b]$;

(ii) $\gamma(t) = e^{it}, \quad t \in [0, \pi]$.

Figs. 22 and 23 show the paths of the arcs γ.

Solution

(i)

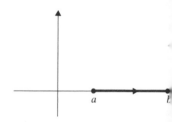

Fig. 22

$$\int_\gamma z^2 \, dz = \int_a^b t^2 \, dt, \quad \text{since } [\gamma(t)]^2 = t^2 \text{ and } \gamma'(t) = 1,$$

$$= \tfrac{1}{3}(b^3 - a^3).$$

Of course in this example we are integrating $z \longrightarrow z^2$ over the interval $[a, b]$ of $\mathbf{R}$, and it would have been very worrying if we had obtained a different answer.

(ii)

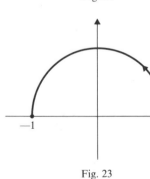

$$\int_\gamma z^2 \, dz = \int_0^\pi e^{2it} \cdot ie^{it} \, dt, \quad \text{since } [\gamma(t)]^2 = (e^{it})^2 = e^{2it} \text{ and } \gamma'(t) = ie^{it},$$

$$= i \int_0^\pi e^{3it} dt$$

$$= i \frac{1}{3i} e^{3it}\Big|_0^\pi$$

$$= \tfrac{1}{3}(e^{3i\pi} - 1) = -\tfrac{2}{3}.$$

Fig. 23

(This is the example we discussed in the Introduction.)

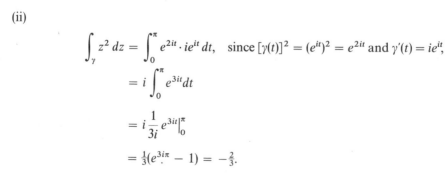

30

Example 2

Evaluate $\int_{\gamma} (z^2 + z)dz$ where $\gamma(t) = t + i, t \in [0, 1]$ (so that the path of γ is the line segment $[i, 1 + i]$).

Solution

$$\int_{\gamma} (z^2 + z)\, dz = \int_0^1 [(t + i)^2 + (t + i)]\, dt$$

$$= \int_0^1 (t^2 + 2ti + t - 1 + i)\, dt$$

$$= \left(\frac{t^3}{3} + t^2 i + \frac{t^2}{2} - t + it \right) \bigg|_0^1$$

$$= -\tfrac{1}{6} + 2i.$$

Notice the way you should evaluate $\int_{\gamma} f(z)\, dz$ in a practical case:

replace z by $\gamma(t)$, dz by $\gamma'(t)dt$, and $\int_{\gamma}$ by $\int_a^b$ (where γ has domain $[a, b]$).

Thus we would seem to evaluate $\int_{\gamma} f(z)dz$ by "substituting" $z = \gamma(t)$. But do not get confused with the usual technique of integration by substitution—there we have a choice of substitutions, but here we *must* use $z = \gamma(t)$.

You have probably noticed in these two examples how we used the fact that complex-valued integrals preserve addition and multiplication (by constants). This property transfers immediately to integrals along smooth arcs—it is not worth calling a theorem, so we shall call it a remark. It is proved by reducing it to complex-valued integrals.

Remark

Let γ be a smooth arc, f and g be functions continuous on the path of γ, and α be a complex constant. Then:

(i) $\quad \int_{\gamma} \alpha f = \alpha \int_{\gamma} f,$

(ii) $\quad \int_{\gamma} (f + g) = \int_{\gamma} f + \int_{\gamma} g.$

Self-Assessment Questions

1. Evaluate $\int_{\gamma} z^2\, dz$ where

   (i) $\quad \gamma(t) = t(1 + i), \quad t \in [1, 4]$;

   (ii) $\quad \gamma(t) = t^2(1 + i), \quad t \in [1, 2]$.

2. Evaluate $\int_{\gamma} \frac{1}{z}\, dz$, where

   (i) $\quad \gamma(t) = e^{it}, \quad t \in [0, 2\pi]$;

   (ii) $\quad \gamma(t) = e^{-it}, \quad t \in [0, 2\pi]$.

3. Evaluate $\int_{\gamma} \sin z\, dz$, where $\gamma(t) = it^2, t \in [1, 2]$.

Solutions

1. (i) $\int_\gamma z^2 dz = \int_1^4 [t(1+i)]^2 (1+i) dt$

$$= (1+i)^3 \int_1^4 t^2 dt$$

$$= 42(-1+i).$$

(ii) $\int_\gamma z^2 dz = \int_1^2 [t^2(1+i)]^2 2t(1+i) dt$

$$= 2(1+i)^3 \int_1^2 t^5 dt$$

$$= 42(-1+i).$$

Note that $\gamma(t) = t^2(1+i)$, $t \in [1,2]$, is a proper reparametrization of the path of $\gamma(t) = t(1+i)$, $t \in [1,4]$, and the two integrals are equal.

2. (i) $\int_\gamma \frac{1}{z} dz = \int_0^{2\pi} e^{-it} \cdot ie^{it} dt = \int_0^{2\pi} i \, dt = 2\pi i.$

(ii) $\int_\gamma \frac{1}{z} dz = \int_0^{2\pi} e^{it} \cdot (-ie^{-it}) dt = \int_0^{2\pi} -i \, dt = -2\pi i.$

3. $\int_\gamma \sin z \, dz = \int_1^2 \sin it^2 \cdot (2it) dt$

$$= -\cos it^2 \Big|_1^2$$

$$= -\cos 4i + \cos i$$

$$= \cosh 1 - \cosh 4.$$

It may have struck you, especially after working Self-Assessment Question 3, that the Fundamental Theorem for complex-valued integrals might transfer to integrals along a smooth arc. This is the case: here is the precise result.

Theorem 6 (The Fundamental Theorem for Integrals along Arcs)

Let f be analytic on a region R, and $\gamma : [a, b] \longrightarrow R$ be a smooth arc in R. Suppose that f' is continuous on R. Then

$$\int_\gamma f'(z) dz = f(\gamma(b)) - f(\gamma(a)).$$

Proof

$$\int_\gamma f'(z) dz = \int_a^b f'(\gamma(t)) \gamma'(t) dt$$

$$= \int_a^b (f \circ \gamma)'(t) dt, \quad \text{by the Chain Rule,}$$

$$= (f \circ \gamma)(b) - (f \circ \gamma)(a), \quad \text{by Theorem 4,}$$

$$= f(\gamma(b)) - f(\gamma(a)). \quad \blacksquare$$

There is an interesting consequence of this result for closed smooth arcs, which is the precursor of many remarkable results in *Unit 5, Cauchy's Theorem I*.

Corollary

Let $\gamma : [a, b] \longrightarrow R$ be a closed smooth arc in a region R, and g be a function continuous on the path of γ. Suppose that there is a function f analytic on R such that $f' = g$ on R. Then $\int_\gamma g$ exists and $\int_\gamma g = 0.$

Proof

$$\int_\gamma g(z)dz = \int_\gamma f'(z)dz$$

$$= f(\gamma(b)) - f(\gamma(a))$$

$$= 0, \quad \text{since } \gamma(b) = \gamma(a), \quad \gamma \text{ being closed.} \quad \blacksquare$$

The Fundamental Theorem is a very good way of evaluating certain integrals along arcs; however, most integrals cannot be evaluated in this way or even by the method used earlier. In fact, much of complex analysis is devoted to finding other ways of evaluating integrals along arcs (as you will see later in the course). In addition, we can often obtain useful information about an integral without evaluating it: Section 4.9 of this unit deals with one way. For such reasons, the problems section which follows deals mostly with general properties of integrals along arcs, rather than with the evaluation of specific integrals.

Summary

In this section we have defined the integral of a continuous function along a smooth arc, discussed some of its elementary properties, and proved the Fundamental Theorem for such integrals. You have also learnt a useful technique: evaluation of these integrals using the Fundamental Theorem.

Self-Assessment Questions

4. Express $\int_\gamma f$ in the form $\int_a^b f(\gamma(t))\gamma'(t)dt$ in each of the following cases.

    (i) $f(z) = z^3$; $\gamma(t) = t + 1, t \in [1, 2]$.

    (ii) $f(z) = \log z$; $\gamma(t) = 2 + e^{it}, t \in [-\pi, \pi]$.

    (Do *not* evaluate the integrals.)

5. Let f be an analytic function with continuous derivative on a region R and γ be a closed smooth arc with path in R. What can you say about $\int_\gamma f'(z)dz$?

6. Let $\gamma(t) = \alpha(1 - t) + \beta t, \ t \in [0, 1]$, and $\alpha \neq \beta$. Evaluate, where k is a complex constant,

    (i) $\int_\gamma k \, dz$;

    (ii) $\int_\gamma (z - k)dz$.

33

Solutions

4.   (i) $\displaystyle\int_1^2 (t + i)^3 dt.$

     (ii) $\displaystyle\int_{-\pi}^{\pi} ie^{it}\mathrm{Log}\,(2 + e^{it})\,dt.$

5. $\displaystyle\int_\gamma f'(z)dz = 0$, by the Corollary to the Fundamental Theorem (Theorem 6).

6.   (i) $\displaystyle\int_\gamma k\,dz = k(\gamma(1) - \gamma(0)) = k(\beta - \alpha).$

     (ii) $\displaystyle\int_\gamma (z - k)dz = \tfrac{1}{2}(z - k)^2 \Big|_\alpha^\beta$

$$= \tfrac{1}{2}[(\beta - k)^2 - (\alpha - k)^2]$$
$$= \tfrac{1}{2}[\beta^2 - \alpha^2 - 2k(\beta - \alpha)].$$

4.6 PROBLEMS

1. Evaluate $\int_\gamma z^n dz$ where $\gamma(t) = re^{it}$, $t \in [0, 2\pi]$, and n is an integer.

2. Prove that there is no function f analytic on the region $R = \{z \in \mathbf{C} : z \neq 0\}$ such that $f'(z) = 1/z$, $z \in R$.

   (*Hint*: Use the corollary to the Fundamental Theorem for arcs.)

3. Here is a general method of integration by substitution for integrals along an arc.

   Let h be a function analytic on R with h' continuous and non-zero on R, and $\gamma : [a, b] \longrightarrow R$ be a smooth arc in R.

   (a) Show that $h \circ \gamma$ is a smooth arc.

   (b) Let f be a function continuous on the path of $h \circ \gamma$. Prove that

   $$\int_{h \circ \gamma} f(w)dw = \int_\gamma f(h(z))h'(z)dz.$$

4. Let γ be a smooth arc with domain $[a, b]$ and f be a function continuous on the path of γ. Prove that

   $$\int_{\tilde{\gamma}} f(z)dz = -\int_\gamma f(z)dz.$$

   where $\tilde{\gamma}$ is the reverse of γ. (See Self-Assessment Question 8 of Section 4.7 for a related result.)

5. Let γ be an arc with domain $[a, b]$, and α be some complex number. Let $\gamma + \alpha$ be the arc defined in the obvious way:

   $$(\gamma + \alpha)(t) = \gamma(t) + \alpha, \qquad t \in [a, b].$$

   (a) Draw a diagram showing the relationship between the paths of γ and $\gamma + \alpha$.

   (b) Let γ be a smooth arc, and f a function continuous on the path of $\gamma + \alpha$. Prove that

   $$\int_{\gamma + \alpha} f(z)dz = \int_\gamma f(z + \alpha)dz.$$

6. Let $\gamma_n(t) = \alpha + re^{nit}$, $t \in [0, 2\pi]$, where n is a non-zero integer. Show that $\dfrac{1}{2\pi i} \displaystyle\int_{\gamma_n} \dfrac{dz}{z - \alpha}$ is an integer.

Problem 6 has dealt with a special case of very useful notion. Given a smooth arc γ and a point α not on the path of γ we define the **winding number of γ about** α, written Wnd(γ, α), to be $\dfrac{1}{2\pi i} \displaystyle\int_\gamma \dfrac{dz}{z - \alpha}$ (which is equal to $\dfrac{1}{2\pi i} \displaystyle\int_a^b \dfrac{\gamma'(t)}{\gamma(t) - \alpha} dt$ if γ has

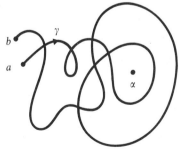

Fig. 24

domain $[a, b]$). As you will see, part (b) of Problem 7, applied to $\gamma - \alpha$, tells us that:

> the winding number of a *closed* smooth arc γ about any point α not on it is an integer.

In other words, the winding number counts how many times the closed arc γ winds round the point α.

35

7. Let γ be a smooth arc with domain $[a, b]$ such that $\gamma \neq 0$ on $[a, b]$. Let
$$l(x) = \int_a^x \frac{\gamma'(t)}{\gamma(t)} dt, \, a \leqslant x \leqslant b.$$

(a) By considering the real and imaginary parts of derivatives of the complex function $m(x) = e^{-l(x)}\gamma(x)$, show that $e^{l(x)} = d\gamma(x)$ for some complex constant d.

(b) If γ is closed, show that $\dfrac{1}{2\pi i} \displaystyle\int_a^b \frac{\gamma'(t)}{\gamma(t)} dt$ is an integer.

Solutions

1.
$$\int_\gamma z^n dz = \int_0^{2\pi} r^n e^{int} \cdot rie^{it} dt$$
$$= r^{n+1} i \int_0^{2\pi} e^{i(n+1)t} dt.$$

If $n = -1$,
$$\int_\gamma z^n dz = i \int_0^{2\pi} 1 dt$$
$$= 2\pi i.$$

If $n \neq -1$,
$$\int_\gamma z^n dz = r^{n+1} i \frac{1}{i(n+1)} (e^{2\pi i(n+1)} - 1)$$
$$= 0.$$

2. Suppose that such an f exists. Consider the arc γ given by $\gamma(t) = e^{it}$, $t \in [0, 2\pi]$ (its path is the circle $|z| = 1$). Then by the corollary to the Fundamental Theorem, since $z \to 1/z$ is continuous on $R = \{z \in \mathbf{C} : z \neq 0\}$,
$$\int_\gamma \frac{1}{z} dz = 0.$$

But by Self-Assessment Question 2 of Section 4.5, $\displaystyle\int_\gamma \frac{1}{z} dz = 2\pi i \neq 0$. Hence we have a contradiction and so such an f does not exist.

(This problem indicates why we cannot define a branch of the logarithm on the region R.)

3. (a) $(h \circ \gamma)'(t) = h'(\gamma(t))\gamma'(t)$ by the Chain Rule. Thus $(h \circ \gamma)'$ exists, is continuous (since h', γ and γ' are), and is non-zero (since h' and γ' are). Hence $h \circ \gamma$ is a smooth arc with domain $[a, b]$.

(b)
$$\int_{h \circ \gamma} f(w) dw = \int_a^b f((h \circ \gamma)(t))(h \circ \gamma)'(t) dt$$
$$= \int_a^b f(h(\gamma(t)))h'(\gamma(t))\gamma'(t) dt$$
$$= \int_a^b g(\gamma(t))\gamma'(t) dt, \quad \text{where } g = (f \circ h) \cdot h',$$
$$= \int_\gamma g(z) dz, \quad \text{by definition,}$$
$$= \int_\gamma f(h(z))h'(z) dz.$$

4. $\displaystyle \int_{\bar{\gamma}} f(z)dz = \int_a^b f(\tilde{\gamma}(t))\tilde{\gamma}'(t)dt$

$\displaystyle = \int_a^b f(\gamma(a+b-t))\cdot(-\gamma'(a+b-t))dt$

$\displaystyle = -\int_a^b \phi(a+b-t)dt, \quad \text{where } \phi = (f\circ\gamma)\cdot\gamma',$

$\displaystyle = \int_b^a \phi(u)du, \quad \text{using the substitution } u = a+b-t, \, du = -dt,$

$\displaystyle = -\int_a^b \phi(u)du$

$\displaystyle = -\int_\gamma f(z)dz.$

5. (a)

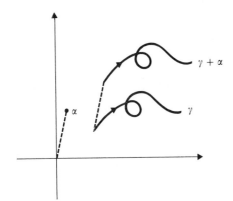

Fig. 25

(b) Use Problem 3(b) with $h(z) = z + \alpha$, for then

$$h\circ\gamma = \gamma + \alpha.$$

6. $\displaystyle \frac{1}{2\pi i}\int_{\gamma_n}\frac{dz}{z-\alpha} = \frac{1}{2\pi i}\int_0^{2\pi}\frac{1}{re^{nit}}\cdot rnie^{nit}dt$

$\displaystyle = \frac{1}{2\pi i}\int_0^{2\pi} ni\, dt$

$\displaystyle = n.$

7. (a) $m'(x) = e^{-l(x)}\gamma'(x) - e^{-l(x)}l'(x)\gamma(x)$

$\displaystyle = e^{-l(x)}\gamma'(x) - e^{-l(x)}\gamma'(x), \quad \text{since } l'(x) = \gamma'(x)/\gamma(x),$

$\displaystyle = 0.$

Hence $\text{Re}(m'(x)) = 0$ and $\text{Im}(m'(x)) = 0$, and so there is a real number h such that $\text{Re}(m(x)) = h$ for $x \in [a, b]$ and a real number k such that $\text{Im}(m(x)) = k$ for $x \in [a, b]$. Therefore there is a complex number $c = h + ik$ such that $m(x) = c$ for $x \in [a, b]$. In other words,

$$e^{-l(x)}\gamma(x) = c.$$

that is, if $d = 1/c$,

$$e^{-l(x)} = d\gamma(x), \quad a \leqslant x \leqslant b.$$

(Note that $c \neq 0$ since $\gamma(x)$ and $e^{-l(x)}$ are non-zero.)

(b) If γ is closed, $\gamma(a) = \gamma(b)$. Thus

$$e^{l(a)} = e^{l(b)}, \quad \text{by part (a).}$$

Hence $e^{l(b)} = 1$ (since $l(a) = 0$) and so $l(b) = 2n\pi i$ for some integer n, by the properties of exp. But $l(b) = \displaystyle\int_a^b\frac{\gamma'(t)}{\gamma(t)}dt$ and hence $\displaystyle\frac{1}{2\pi i}\int_a^b\frac{\gamma'(t)}{\gamma(t)}dt$ is an integer.

4.7 CONTOUR INTEGRALS

In the last reading section and the problems afterwards, we dealt with some of the *theory* of integrals along smooth arcs. But when it comes to using them in *practice*, that is, how they crop up in various parts of complex analysis, we find that we are not usually given the arc, but only its directed path. In this section we discuss how to define integrals along directed paths.

*   *   *   *   *   *   *   *

In Self-Assessment Question 1 of Section 4.5 we observed that for a particular function its integrals along two different arcs, one of which was a proper reparametrization of the other, were equal. So one is naturally led to ask how integrals along arcs are affected by proper reparametrization. The next theorem provides the answer: they aren't.

Theorem 7 (The Reparametrization Theorem for Integrals along Arcs)

Let γ_1 and γ_2 be two smooth arcs (with domains $[a, b]$ and $[c, d]$ respectively) such that γ_1 is a proper reparametrization of the path of γ_2 by $\theta : [a, b] \longrightarrow [c, d]$. Let f be any function continuous on the path of γ_1. Then

$$\int_{\gamma_1} f(z)dz = \int_{\gamma_2} f(z)dz.$$

Proof

$$\int_{\gamma_2} f(z)dz = \int_c^d f(\gamma_2(t))\gamma_2'(t)dt$$

$$= \int_c^d f(\gamma_1(\theta(t)))\gamma_1'(\theta(t))\theta'(t)dt,$$

since $\gamma_2(t) = \gamma_1(\theta(t))$ and $\gamma_2'(t) = \gamma_1'(\theta(t))\theta'(t)$,

$$= \int_a^b f(\gamma_1(s))\gamma_1'(s)ds,$$

using the substitution $s = \theta(t)$, $ds = \theta'(t)dt$,

$$= \int_{\gamma_1} f(z)dz. \quad \blacksquare$$

For simple smooth arcs, we can deduce the following corollary.

Corollary

Let γ_1 and γ_2 be two simple smooth arcs with the same directed path Γ, and let f be a function continuous on Γ. Then

$$\int_{\gamma_1} f = \int_{\gamma_2} f.$$

Proof

γ_1 is a proper reparametrization of γ_2 by some function θ (see page 23). The result then follows by applying Theorem 7. $\quad \blacksquare$

This corollary justifies the following notation. Let Γ be some directed path of a simple smooth arc, and f be continuous on Γ. Then $\int_\Gamma f$ is defined to be $\int_\gamma f$

where γ is some simple smooth arc with path Γ, described in the right direction.

A similar uniqueness theorem holds when γ_1 and γ_2 are simple-closed smooth arcs with the same directed path, and so again $\int_\Gamma f$ is well-defined when Γ is the directed path of a simple-closed smooth arc. In this case it is also true that $\int_\Gamma f$ is independent of the initial point. (The proof of this result is relatively easy, but tedious: we therefore omit it.)

Self-Assessment Questions

1. Evaluate $\int_\Gamma z^2 dz$ where

    (i) Γ is the line segment $[a, b]$;

    (ii) Γ is the semi-circle $\{z : |z| = 1, \operatorname{Im} z \geqslant 0\}$ from 1 to -1.

2. Evaluate $\int_\Gamma \frac{1}{z} dz$ where

    (i) Γ is the circle $|z| = 1$ with positive orientation;

    (ii) Γ is the circle $|z| = 1$ with negative orientation.

Solutions

1. (i)   Let $\gamma(t) = t, t \in [a, b]$. Then

    $$\int_\Gamma z^2 \, dz = \int_\gamma z^2 \, dz$$
    $$= \tfrac{1}{3}(b^3 - a^3), \quad \text{by Example 1(i) of Section 4.5.}$$

    (ii) Let $\gamma(t) = e^{it}, t \in [0, \pi]$. Then

    $$\int_\Gamma z^2 \, dz = \int_\gamma z^2 \, dz$$
    $$= -\tfrac{2}{3}, \quad \text{by Example 1(ii) of Section 4.5.}$$

2. (i)   Let $\gamma(t) = e^{it}, t \in [0, 2\pi]$. Then

    $$\int_\Gamma \frac{1}{z} dz = 2\pi i, \quad \text{by Self-Assessment Question 2(i) of Section 4.5.}$$

    (ii) Let $\gamma(t) = e^{-it}, t \in [0, 2\pi]$. Then

    $$\int_\Gamma \frac{1}{z} dz = -2\pi i, \quad \text{by Self-Assessment Question 2(ii) of Section 4.5.}$$

You may have noticed that in Self-Assessment Question 1 above we essentially used the Fundamental Theorem for integrals along arcs. This opens up the prospect that in such cases we need not explicitly calculate the smooth arc, since all the information we need is the endpoints, which are given by the path.

In view of this, try the next two problems. (If you are short of time, you should treat them as examples.)

Problems

1. Write down and prove a suitable version for directed paths of the Fundamental Theorem for integrals along an arc (Theorem 6 on page 32).

2. Write down suitable versions for directed paths of the Remark on page 31.

Solutions

1. We shall prove the following:

   Let f be analytic on a region R and let Γ be a directed path (of some simple smooth arc) contained in R. Suppose also that f is continuous on R. Then

   $$\int_{\Gamma} f'(z)dz = f(\beta) - f(\alpha)$$

   where α is the initial and β the final point of Γ.

   To prove this, let $\gamma:[a, b] \longrightarrow R$ be a smooth arc with directed path Γ. Then

   $$\int_{\Gamma} f'(z)dz = \int_{\gamma} f'(z)dz$$
   $$= f(\gamma(b)) - f(\gamma(a)), \quad \text{by Theorem 6,}$$
   $$= f(\beta) - f(\alpha).$$

Notice that Self-Assessment Question 1 now becomes very easy because $\dfrac{d}{dz}\left(\dfrac{z^3}{3}\right) = z^2$, and, therefore, for part (i)

$$\int_{\Gamma} z^2 dz = \frac{z^3}{3}\bigg|_a^b = \tfrac{1}{3}(b^3 - a^3),$$

and for part (ii)

$$\int_{\Gamma} z^2 dz = \frac{z^3}{3}\bigg|_1^{-1} = -\tfrac{2}{3}.$$

Notice, however, that Self-Assessment Question 2 does not yield to this method because there is no region R containing Γ and function L such that $L'(z) = 1/z$ on R. (See Problem 2 of Section 4.6.)

2. Let Γ be a directed path, f and g be functions continuous on Γ, and α be a complex constant. Then

   (i) $\displaystyle\int_{\Gamma} \alpha f = \alpha \int_{\Gamma} f$;

   (ii) $\displaystyle\int_{\Gamma} (f + g) = \int_{\Gamma} f + \int_{\Gamma} g.$

(We have not dignified these two results on paths with the status of a theorem, since we shall prove more general results later.)

By joining up smooth arcs we can integrate over arcs which have "corners"— such as polygonal lines. This is what we now want to discuss.

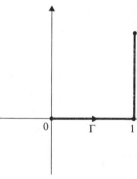

Fig. 26

Consider the path Γ shown in Fig. 26. There is *no* smooth arc which has Γ as its path (because of the corner at 1). But it is fairly clear how we integrate, say, the function $z \longrightarrow z^2$ along Γ: we just define $\displaystyle\int_{\Gamma} z^2\, dz$ to be

$$\int_{[0,1]} z^2\, dz + \int_{[1,1+i]} z^2\, dz,$$

which equals

$$\tfrac{1}{3}(1^3 - 0^3) + \tfrac{1}{3}[(1 + i)^3 - 1^3] = \tfrac{1}{3}(1 + i)^3 = -\tfrac{2}{3}(1 - i).$$

This example motivates our next definitions.

40

Definition

A **contour** is a finite sequence $(\Gamma_0, \ldots, \Gamma_{n-1})$ of directed paths of simple or simple-closed smooth arcs such that for each $i = 1, \ldots, n-1$, the initial point of Γ_i is the final point of Γ_{i-1}.

For contours we shall again use the symbol Γ. This will cause no confusion with directed paths of smooth arcs for reasons which will rapidly become apparent. As for directed paths, we shall also use Γ to denote the set which is the path (in an obvious sense) of Γ.

We extend the terms "simple" and "closed" to contours in the obvious way.

simple contour   contour, not simple   closed contour

Fig. 27

A simple-closed contour is presumed to have positive direction or orientation, unless otherwise stated. (See Example 6, page 24.)

Definition

Let Γ be a contour and f be a function continuous on Γ. Then the **integral of f along Γ**, written $\int_\Gamma f$ or $\int_\Gamma f(z)\, dz$, is defined by

$$\int_\Gamma f = \int_{\Gamma_1} f + \cdots + \int_{\Gamma_n} f,$$

where $\Gamma = (\Gamma_1, \ldots, \Gamma_n)$ in the sense of the previous definition. Such an integral is called a **contour integral**.

It should be fairly clear that for a given contour Γ, $\int_\Gamma f$ is uniquely determined, and if Γ is closed then it does not matter from which point of Γ we choose to start our integration.

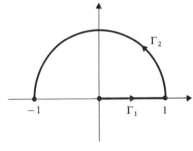

Fig. 28

Self-Assessment Questions

3. Evaluate $\int_\Gamma z\, dz$ where Γ is the polygonal line $[0, 1, i]$.

4. Evaluate $\int_\Gamma 3(z^2 + 1)\, dz$ where Γ is the contour shown in Fig. 28.

Solutions

3. $$\int_\Gamma z\, dz = \int_{[0,1]} z\, dz + \int_{[1,i]} z\, dz$$
    $$= \tfrac{1}{2}z^2\big|_0^1 + \tfrac{1}{2}z^2\big|_1^i$$
    $$= \tfrac{1}{2}z^2\big|_0^i = -\tfrac{1}{2}.$$

4. $\int_\Gamma 3(z^2 + 1)\,dz = \int_{\Gamma_1} 3(z^2 + 1)\,dz + \int_{\Gamma_2} 3(z^2 + 1)\,dz$

$$= 3\int_{\Gamma_1} z^2\,dz + 3\int_{\Gamma_1} 1\,dz + 3\int_{\Gamma_2} z^2\,dz + 3\int_{\Gamma_2} 1\,dz$$

$$= 3\int_\Gamma z^2\,dz + 3\int_\Gamma 1\,dz$$

$$= 3(\tfrac{1}{3}z^3)\big|_0^{-1} + 3(z)\big|_0^{-1}$$

$$= -4.$$

The method of Self-Assessment Question 3 readily generalizes, and we get the following important result for the evaluation of contour integrals.

Theorem 8 (The Fundamental Theorem for Contour Integrals)

Let f be analytic on a region R, and Γ be a contour in R. Suppose also that f' is continuous on R. Then

$$\int_\Gamma f'(z)\,dz = f(\beta) - f(\alpha),$$

where α is the initial and β the final point of Γ. Furthermore, if Γ is closed, then

$$\int_\Gamma f'(z)\,dz = 0.$$

Proof

Let $\Gamma = (\Gamma_1, \ldots, \Gamma_n)$ and $\gamma_i : [a_i, b_i] \longrightarrow R$ be a smooth arc with directed path Γ_i, for each i. Then

$$\int_\Gamma f'(z)\,dz = \int_{\Gamma_1} f' + \cdots + \int_{\Gamma_n} f'$$

$$= \int_{\gamma_1} f' + \cdots + \int_{\gamma_n} f'$$

$$= f(\gamma_1(b_1)) - f(\gamma_1(a_1)) + \cdots + f(\gamma_n(b_n)) - f(\gamma_n(a_n)),$$

$$\text{by Theorem 6,}$$

$$= f(\gamma_n(b_n)) - f(\gamma_1(a_1)),$$

$$\text{since } \gamma_i(a_i) = \gamma_{i-1}(b_i) \text{ for each } i = 1, \ldots, n,$$

$$= f(\beta) - f(\alpha).$$

If $\beta = \alpha$ (that is, Γ is closed) then $\int_\Gamma f' = 0$. ∎

Similarly, the technique of Self-Assessment Question 4 generalizes; we obtain the following result.

Theorem 9

Let Γ be a contour in a region R, f and g be functions continuous on R, and α be a complex constant. Then

(i) $\displaystyle \int_\Gamma \alpha f = \alpha \int_\Gamma f;$

(ii) $\displaystyle \int_\Gamma (f + g) = \int_\Gamma f + \int_\Gamma g.$

Proof

Let $\Gamma = (\Gamma_1, \ldots, \Gamma_n)$. For variety we shall prove the result of (i) using $\sum$ notation and Problem 2 (page 39).

(i) $\quad \displaystyle\int_\Gamma \alpha f = \sum_{r=1}^n \int_{\Gamma_r} \alpha f = \sum_{r=1}^n \alpha \int_{\Gamma_r} f = \alpha \sum_{r=1}^n \int_{\Gamma_r} f = \alpha \int_\Gamma f.$

(ii) We leave this to you. ∎

Self-Assessment Question 5

Evaluate $\displaystyle\int_\Gamma (z^2 + \sinh z)\, dz$ where Γ is the square contour $[0, 1, 1 + i, i, 0]$.

Solution

$$\int_\Gamma (z^2 + \sinh z)\, dz = \int_\Gamma z^2\, dz + \int_\Gamma \sinh z\, dz, \quad \text{by Theorem 9,}$$

$$= 0 + 0, \quad \text{by Theorem 8, since } \frac{d}{dz}(\tfrac{1}{3}z^3) = z^2 \text{ and}$$

$$\cosh' = \sinh.$$

There is a more convenient notation for contours than the one introduced in the definition, which ties up rather better with the definition of contour integrals. Given contours $\Gamma = (\Gamma_1, \ldots, \Gamma_m)$ and $\Delta = (\Delta_1, \ldots, \Delta_n)$ such that the initial point of Δ is the final point of Γ, we define the *sum* $\Gamma + \Delta$ of Γ and Δ to be the contour

$$(\Gamma_1, \ldots, \Gamma_m, \Delta_1, \ldots, \Delta_n).$$

Note that the initial point of Δ_1 is the final point of Γ_m. Intuitively, we just place Δ after Γ.

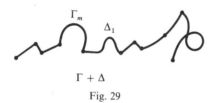

$\Gamma + \Delta$

Fig. 29

It is then clear that if f is continuous on $\Gamma + \Delta$ then

$$\int_{\Gamma + \Delta} f = \int_\Gamma f + \int_\Delta f.$$

Using the sum notation we can write any contour $\Gamma = (\Gamma_1, \ldots, \Gamma_n)$ in the form

$$\Gamma = \Gamma_1 + \ldots + \Gamma_n,$$

since each directed path Γ_i is a contour.

If Γ is the directed path of some simple, or simple-closed, smooth arc, we define the **reverse path of** Γ, denoted by $-\Gamma$, to be the directed path of the reverse arc $\tilde{\gamma}$ of any smooth arc γ with path Γ. The **reverse of a contour** Γ (defined in the obvious way) is also denoted by $-\Gamma$. (See Self-Assessment Question 8.)

Summary

In this section we have defined the integral of a function along a contour, and we have described some important properties of this integral.

Self-Assessment Questions

6. Let $\gamma_1(t) = it$, $t \in [1, 4]$, and $\gamma_2(t) = i(t - 1)^2$, $t \in [2, 3]$. What can you say about the relation between $\int_{\gamma_1} f$ and $\int_{\gamma_2} f$ for a continuous function f? Give your reason.

7. Let p be a polynomial and Γ be a contour. What can you say about $\int_\Gamma p(z)dz$ if

   (i) Γ is closed,

   (ii) Γ is not closed?

8. Let f be continuous on a contour Γ. Prove that

$$\int_{-\Gamma} f = -\int_\Gamma f.$$

Solutions

6. $\int_{\gamma_1} f = \int_{\gamma_2} f$ by the Reparametrization Theorem (Theorem 7).

7. (i)   If Γ is closed then $\int_\Gamma p(z)dz = 0$, by the Fundamental Theorem (Theorem 8).

   (ii) If Γ is a contour from α to β then $\int_\Gamma p(z)dz$ depends only on α and β and not on the other points of Γ, by the Fundamental Theorem.

8. Let $\Gamma = (\Gamma_1, \ldots, \Gamma_n)$ be a contour. Then the reverse contour $-\Gamma$ of Γ is

$$-\Gamma = (-\Gamma_n, -\Gamma_{n-1}, \ldots, -\Gamma_1).$$

Fig. 30

Let f be continuous on Γ. Then

$$\int_{-\Gamma} f = \sum_{i=1}^{n} \int_{-\Gamma_i} f = \sum_{i=1}^{n} -\int_{\Gamma_i} f = -\int_\Gamma f.$$

4.8 PROBLEMS

1. Express each of the following integrals $\int_\Gamma f(z)dz$ (where Γ is a directed path) in the form

$$\int_{t_0}^{t_1} f(\gamma(t))\gamma'(t)dt.$$

(i) $\int_\Gamma \mathrm{Log}\, z\, dz$, where Γ is the right-hand half of the circle $|z| = 1$, from $-i$ to i.

(ii) $\int_\Gamma \sin z\, dz$, where Γ is the line segment $[i, 1]$.

2. Evaluate the following contour integrals:

(i) $\int_\Gamma e^{-\pi z}dz$, where Γ is any contour from $-i$ to i;

(ii) $\int_\Gamma (z-1)^3\, dz$, where Γ is any contour from 2 to $2i + 1$;

(iii) $\int_\Gamma \sin z\, dz$, where Γ is any contour from i to 1.

3. Use the Fundamental Theorem for contour integrals to prove the following result:

If f is analytic on a region R and $f' = 0$ on R, then f is constant on R.

Solutions

1. (i) Γ is the path of γ, where $\gamma(t) = e^{it}$, $t \in [-\pi/2, \pi/2]$. Thus

$$\int_\Gamma \text{Log } z \, dz = \int_{-\pi/2}^{\pi/2} \text{Log } e^{it} \cdot ie^{it} \, dt$$

$$= -\int_{-\pi/2}^{\pi/2} te^{it} dt, \quad \text{since Log } e^{it} = it.$$

(ii) Γ is the path of γ, where $\gamma(t) = t + i(1-t)$, $t \in [0, 1]$. Thus

$$\int_\Gamma \sin z \, dz = \int_0^1 \sin(t + i(1-t)) \cdot (1-i) dt.$$

2. We apply the Fundamental Theorem for contour integrals.

(i) $\displaystyle\int_\Gamma e^{-\pi z} dz = -\frac{1}{\pi} e^{-\pi z} \Big|_{-i}^{i} = -\frac{1}{\pi}[(-1) - (-1)] = 0.$

(ii) $\displaystyle\int_\Gamma (z-1)^3 dz = \tfrac{1}{4}(z-1)^4 \Big|_2^{2i+1} = \tfrac{1}{4}(16i^4 - 1) = \tfrac{15}{4}.$

(iii) $\displaystyle\int_\Gamma \sin z \, dz = -\cos z \Big|_i^1 = -\cos 1 - (-\cos i) = \cosh 1 - \cos 1.$

3. Let $a, b \in R$. We must show that $f(a) = f(b)$. If $[a, b] \in R$, and γ is the smooth arc given by $\gamma(t) = a(1-t) + bt$, $t \in [0, 1]$, then

$$f(b) - f(a) = \int_\gamma f'(z) dz$$

$$= \int_0^1 f'(a(1-t) + bt) \, dt$$

$$= \int_0^1 0 \, dt = 0, \quad \text{by the Fundamental Theorem.}$$

In general, there is a polygonal contour Γ from a to b (since R is polygonally-connected). Now restrict Γ so that it intersects itself only at the initial and final points of the Γ_i. Then, by the Fundamental Theorem,

$$f(b) - f(a) = \int_\Gamma f'$$

$$= \int_{\Gamma_1} f' + \ldots + \int_{\Gamma_n} f', \quad \text{where } \Gamma = (\Gamma_1, \ldots, \Gamma_n).$$

$$= 0 + \ldots + 0, \quad \text{by the previous remark,}$$

$$= 0.$$

4.9 THE ESTIMATION THEOREM

Given a particular contour Γ and a function f continuous on Γ, it is very often impossible to evaluate $\int_\Gamma f$ without resorting to numerical techniques. However we can derive a result which gives an upper bound for the *modulus* of $\int_\Gamma f$.

This result is not hard to prove but it is of immense theoretical importance: indeed, most of the major results of complex analysis involve at least one use of it.

In order to prove this result we shall need the result of Problem 7 of Section 4.2, in which we asked you to prove that if ϕ is a complex-valued function whose domain contains $[a, b]$, then

$$\left| \int_a^b \phi(t)dt \right| \leqslant \int_a^b |\phi(t)|dt.$$

We shall also require the notion of *length of a contour*. If $\Gamma = \Gamma_1 + \ldots + \Gamma_n$ is a contour, then the length of Γ is the sum of the lengths of a set of simple smooth arcs γ_i whose paths are Γ_i, $i = 1, 2, \ldots, n$. (The arcs γ_i, of course, could be simple-closed smooth arcs: any statements in this section about simple smooth arcs apply also to simple-closed smooth arcs.)

Theorem 10 (The Estimation Theorem)

Let Γ be a contour and f a function continuous on Γ. Then

$$\left| \int_\Gamma f \right| \leqslant ML,$$

where L is the length of Γ and M is an upper bound of $|f|$ on Γ.

Proof

It is clearly sufficient to prove this result when Γ is the directed path of a simple smooth arc, since the general result for a contour $\Gamma = (\Gamma_1, \ldots, \Gamma_n)$ can be obtained by adding the result for each Γ_i.

So let Γ be the directed path of some simple smooth arc γ, with domain $[a, b]$, say. Then

$$\int_\Gamma f = \int_a^b f(\gamma(t))\gamma'(t)dt.$$

Using the result mentioned above, with $\phi(t) = f(\gamma(t))\gamma'(t)$, we get

$$\left| \int_\Gamma f \right| \leqslant \int_a^b |f(\gamma(t))| \cdot |\gamma'(t)|dt$$

$$\leqslant \int_a^b M|\gamma'(t)|dt, \quad \text{since } |f| \leqslant M \text{ on } \Gamma,$$

$$= M \int_a^b |\gamma'(t)|dt$$

$$= ML, \quad \text{by the definition of } L. \quad \blacksquare$$

Example

Find an upper bound for $\left| \int_C e^z dz \right|$ where C is the circle $|z| = 1$.

Solution

If $|z| = 1$, then

$$|e^z| = e^x |e^{iy}|, \quad \text{where } z = x + iy,$$

$$= e^x \leqslant e, \quad \text{since } x \leqslant 1.$$

Let $M = e$ and $L = 2\pi$ in Theorem 10; thus

$$\left| \int_C e^z \, dz \right| \leqslant ML = 2\pi e.$$

Theorem 10 does not have much *direct* practical usefulness, but when allied with the Residue Theorem (of *Unit 9, Cauchy's Theorem II*) it becomes a highly useful technique for evaluating integrals. This will be discussed in detail in *Unit 10, The Calculus of Residues.*

Self-Assessment Question

Let α and β be such that $|\alpha| \geqslant |\beta|$, and $\Gamma = [\alpha, \beta]$. Use the Estimation Theorem to show that

$$\left| \int_\Gamma z^2 dz \right| \leqslant |\alpha|^2 |\beta - \alpha|.$$

Solution

For z on Γ, $|z| \leqslant |\alpha|$. Let $M = |\alpha|^2$, $L = |\beta - \alpha|$ in the Estimation Theorem. Then

$$\left| \int_\Gamma z^2 dz \right| \leqslant |\alpha|^2 |\beta - \alpha|.$$

Problems

1. Let Γ be the circle $|z| = 1$. Find an upper bound for

    $$\left| \int_\Gamma \frac{dz}{3 + 4z} \right|.$$

2. Let Γ be the semi-circle $\{z : |z| = r, \operatorname{Im} z \geqslant 0\}$ from $-r$ to r, and a be some positive real number.

    (a) Prove that

    $$\left| \int_\Gamma \frac{dz}{z^2 + a^2} \right| \leqslant \frac{\pi r}{r^2 - a^2} \quad \text{if } r > a,$$

    and find the corresponding result for $r < a$.

    (b) What can you say about $\lim\limits_{r \to \infty} \left| \int_\Gamma \frac{dz}{z^2 + a^2} \right|$?

Solutions

1. For $|z| = 1$,

$$\left| \frac{1}{3 + 4z} \right| = \frac{1}{|3 + 4z|}$$

$$\leqslant \frac{1}{4|z| - 3}, \quad \text{using the inequality } |u + v| \geqslant |u| - |v|,$$

$$= 1.$$

Thus

$$\left| \int_{\Gamma} \frac{dz}{3 + 4z} \right| \leqslant 1 \cdot 2\pi = 2\pi, \quad \text{by the Estimation Theorem.}$$

2. (a) The length L of Γ is πr. To find an upper bound for $\left| \dfrac{1}{z^2 + a^2} \right|$ when $|z| = r$ (where $r > a$), we argue as follows:

$$\left| \frac{1}{z^2 + a^2} \right| = \frac{1}{|z^2 + a^2|}$$

$$\leqslant \frac{1}{|z|^2 - a^2}, \quad \text{using the inequality } |u + v| \geqslant |u| - |v|,$$

$$= \frac{1}{r^2 - a^2}.$$

Let $M = 1/(r^2 - a^2)$, Hence by the Estimation Theorem, we get

$$\left| \int_{\Gamma} \frac{dz}{z^2 + a^2} \right| \leqslant \frac{\pi r}{r^2 - a^2}.$$

If $r < a$, the same method yields the upper bound $\dfrac{\pi r}{a^2 - r^2}$.

(b) Since

$$\lim_{r \to \infty} \frac{\pi r}{r^2 + a^2} = \lim_{r \to \infty} \frac{\pi}{r + a^2/r} = 0,$$

and $a < r$ for large enough r,

$$\lim_{r \to \infty} \left| \int_{\Gamma} \frac{dz}{z^2 + a^2} \right| = 0.$$

Unit 5   Cauchy's Theorem I

Conventions

Before working through this text make sure you have read *A Guide to the Course : Complex Analysis*.

References to units of other Open University courses in mathematics take the form:

Unit M100 13, Integration II.

The set book for the course M231, Analysis, is M. Spivak, *Calculus*, paperback edition (W. A. Benjamin/Addison-Wesley, 1973). This is referred to as:

Spivak.

Optional Material

This course has been designed so that it is possible to make minor changes to the content in the light of experience. You should therefore consult the supplementary material to discover which sections of this text are not part of the course in the current academic year.

5.0 INTRODUCTION

In this unit we prove the central result of complex analysis: Cauchy's Theorem. From it will flow a veritable host of surprising results, as you will see throughout this course. Roughly speaking, Cauchy's Theorem says the following: if f is a function analytic on a region R which, in some sense, is "rather like" a disc and Γ is a closed contour in R, then

$$\int_{\Gamma} f(z)\, dz = 0.$$

In fact the proof of the most general form of Cauchy's Theorem is quite difficult, and we shall not discuss it until *Unit 9, Cauchy's Theorem II*. However, many straightforward applications follow from a less general form of Cauchy's Theorem, for the so-called *star regions*. To find out how these arise naturally, we begin Section 5.1 by discussing the Fundamental Theorem of Calculus, which may not strike you as closely related: this leads up to the definition and properties of star regions. In Section 5.3 we prove Cauchy's Theorem via the Antiderivative Theorem. The following two reading sections deal with some of the ways in which Cauchy's Theorem is useful in complex analysis, and we close with a more down-to-earth section on ways of evaluating integrals.

Television

In the third television programme associated with the course we shall illustrate the applications of Cauchy's Formula to the evaluation of integrals.

5.1 STAR REGIONS

We begin by recalling some results from real analysis.

Let f be a real function continuous on the closed interval $[a, b]$ of $\mathbf{R}$. Then

(1) the function $F(x) = \int_a^x f(t)\, dt$ is differentiable on $[a, b]$ and $F'(x) = f(x)$.

From this result two consequences follow, one theoretical, and one more practical. On the theoretical side, we can deduce trivially,

(1a) there is a function F such that $F' = f$ on $[a, b]$.

The function F is often called a *primitive* of f (as in M231, Analysis), but we shall call F an **antiderivative of** f **on** $[a, b]$. On the practical side, we have

(2) given some antiderivative G of f on $[a, b]$, then

$$\int_a^b f(t)\, dt = G(b) - G(a).$$

This is the result which enables us to evaluate integrals. Note that (2) follows from (1) and the fact that, since $F' = G'$, F and G differ only by some constant.

Depending on what book you read, one or more of these results is called the Fundamental Theorem of Calculus. The reason for this ambiguity is probably that the results are so closely related.

Let us now consider whether there are analogues of these results in complex analysis. Certainly there is an analogue of (2), since in *Unit 4, Integration*, the following result was proved. (We have changed the words slightly to fit the present discussion.)

Theorem 8 of Unit 4 (The Fundamental Theorem for Contour Integrals)

Let f be continuous on a region R, and let F be analytic on R, with $F'(z) = f(z)$, $z \in R$. Then, given any contour Γ in R with initial point a and final point b,

$$\int_\Gamma f(z)\, dz = F(b) - F(a).$$

Since (1) has no obvious analogue, let us look at (1a), which is after all, an easy consequence of (1). Given a function f continuous on a region R, is there a function F analytic on R such that $F'(z) = f(z)$, $z \in R$? The answer is: No. For example, it can be shown that no such F exists for $f(z) = |z|$ on any region. (See SAQ 8 of Section 5.5.)

Obviously, then, continuity is not a strong enough assumption in the complex plane. The obvious extra hypothesis is analyticity, and so we are led to pose the following question.

Question

Given a function f analytic on a region R, is there a function F analytic on R such that $F'(z) = f(z)$, $z \in R$?

Unfortunately, the answer is still: No. For example, if $f(z) = 1/z$ and $R = \{z \in \mathbf{C} : z \neq 0\}$, then no such F exists. (See Problem 2 of Section 4.6 in *Unit 4*.) Since analytic functions seem so well behaved, we are led to ask: Could we answer "Yes" to the question if R was some *special* kind of region? To get some idea of what kind of region we should consider we look back to result (1).

Let f be analytic on some region R. By analogy with the real case, we might try to construct a function F such that $F' = f$ by taking some fixed point $z_0 \in R$ and defining $F(z)$ as the integral from z_0 to z of f. But in complex analysis we cannot just specify the endpoints—we must give a contour as well. The most obvious contour is the line segment $[z_0, z]$, and so we could define a function F by

$$F(z) = \int_{[z_0, z]} f.$$

However, for this to *exist* for all $z \in R$, we must have $[z_0, z]$ contained in R for all $z \in R$ (because f may not even be continuous on any set larger than R). This suggests that regions R with the property that $[z_0, z] \subseteq R$ for all $z \in R$, where z_0 is some particular point in R, might repay investigation. Such regions are the topic of the rest of this section. As well as their usefulness in later work (see Section 5.3), they form an interesting topic in their own right, and so we shall study them fairly thoroughly.

Definition

A set S is called a **star set** if there is a point $z_0 \in S$ such that for all $z \in S$ the line segment $[z_0, z]$ is contained in S. The point z_0 is called a **star for** S.

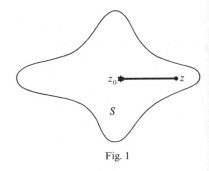

Fig. 1

We can imagine z_0 as a light source: then z_0 is a star for S just if every point of S is illuminated by z_0.

A region which is a star set is called a **star region**. The following remark is very useful.

Remark

Any open set G which is a star set is a star region.

Proof

Let z_0 be a star for G. Given $z, z' \in G$, then $[z_0, z]$ and $[z_0, z']$ lie in G (Fig. 2) and so the polygonal line $[z, z_0, z']$ lies in G. Thus G is polygonally-connected and so a region. ∎

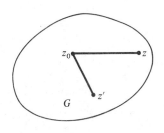

Fig. 2

Thus, having a star is a very special case of polygonal connectedness. You may recall that we have already seen one other special case—in *Unit 2, Continuous Functions*, a set S was called *convex* if given any two points $z, z' \in S$ then $[z, z']$ is contained in S. Clearly, a convex set S is a star set, since in fact every point in it is a star for S: and, conversely, if every point in S is a star for S, then S is convex. These two concepts are even more intertwined than this remark suggests: in order to study star regions one has simultaneously to study convex regions. Each makes up for the deficiencies of the other, as we shall see.

It is often easy to tell whether or not a region is a star region, by looking at a picture of it. Thus the following regions are star regions. In fact they are convex—every point in them is a star.

Fig. 3

55

For non-convex regions, such as those in Fig. 4, we can choose only certain points as stars (those in the darker parts).

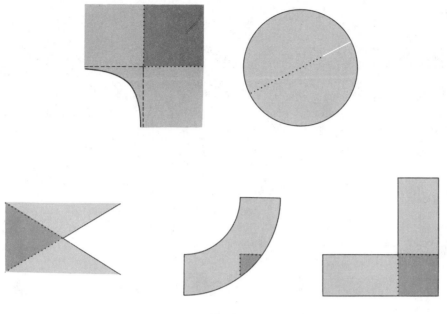

Fig. 4

It is even possible that a star region has only *one* star. One example is given in Fig. 5: there are many others. This example may serve to show one defect of using pictures to determine whether or not a region is a star region. A picture is only an approximation, and since the definition of star set is geometric in nature, a star set is not necessarily preserved under small deformations. For example, if the three cuts in the region in Fig. 5 are deformed so that they do not meet when prolonged, at one point, the region will no longer be a star region.

Fig. 5

Thus, when asked, you should be able to give a rigorous analytic proof that a region is a star region, or at least give sufficient of a geometric argument to show how an analytic proof would be constructed. We now embark on systematically finding a large stock of star regions, rather in the way we proceeded for open sets in *Unit 2*. To begin with, we look at the "simplest" open sets.

Example 1

Show that the following open sets are star regions, and are, in fact, convex:

(i)   any open disc;

(ii) any open half-plane.

Solution

(i)   Let $D = \{z : |z - \alpha| < r\}$, and $z_1, z_2 \in D$ (Fig. 6). If $z \in [z_1, z_2]$ then

$$|z - \alpha| \leqslant \max(|z_1 - \alpha|, |z_2 - \alpha|) < r,$$

and so $z \in D$.

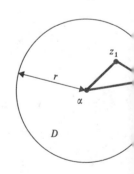

Fig. 6

(ii) To begin with consider the half-plane $R = \{z : \operatorname{Im} z > 0\}$. If $z_1, z_2 \in R$ and $z \in [z_1, z_2]$ then

$$\operatorname{Im} z \geqslant \min (\operatorname{Im} z_1, \operatorname{Im} z_2) > 0,$$

and so $z \in R$.

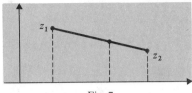

Fig. 7

Any other half-plane is obtained from R by translation and rotation, which clearly preserve convexity (since they preserve line segments).

When it comes to combining star regions, neither intersection nor union preserve the property of being a star region. But it is easy to prove that *convex* regions are preserved under intersection (in fact this is the main reason for studying convex regions in this context) and that the union of convex regions is star (this is the main reason for generalizing convex to star). We shall leave the details as problems for you: the strategy is very similar to that used in Section 2.3 of *Unit 2* for building up a stock of open sets.

Summary

In this section we have introduced the notion of star region, discussed the visual determination of star regions and given some analytic proofs that regions are star.

Self-assessment Questions

1. Write down the definition of "star set".

2. Decide which of the following are star regions.

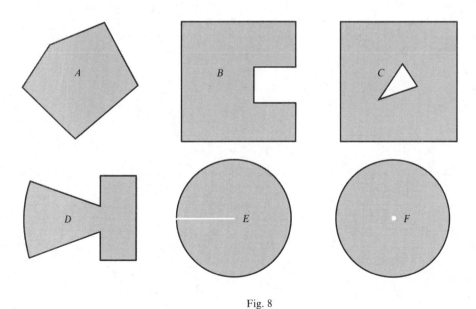

Fig. 8

3. Why is $\mathbf{C}$ convex?

4. Why is the empty set $\emptyset$ convex? Is $\emptyset$ a star set?

Solutions

1. See page 55.

2. A, D and E are star regions: B, C and F are not. In A, D and E we have shaded the set of stars in a darker tone.

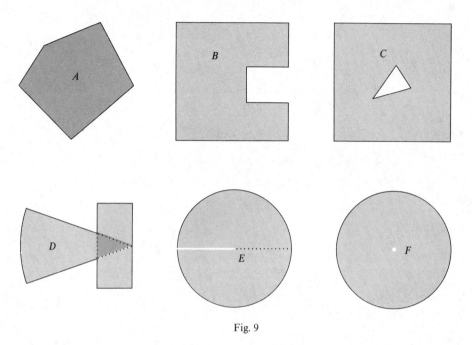

Fig. 9

3. Given any $z, z' \in \mathbf{C}$, then $[z, z'] \subseteq \mathbf{C}$ (since $tz + (1 - t)z' \in \mathbf{C}$, for any $t \in [0, 1]$).

4. We regard $\emptyset$ as convex since there are no points z_1, z_2 in $\emptyset$ such that $[z_1, z_2]$ is not in $\emptyset$. Since a star set must have a star, it is nonempty: therefore $\emptyset$ is not star.

5.2 PROBLEMS

1. First of all, we deal with combining star regions and convex regions.
    (a) Prove that if R_1 and R_2 are convex regions, then $R_1 \cap R_2$ is a convex region. (Note that $R_1 \cup R_2$ need not even be a region.)
    (b) Find two star regions R_1 and R_2 such that $R_1 \cap R_2$ is not a star region.
    (c) Let R_1 and R_2 be star regions, and let z_0 be a star for R_1 and for R_2. Show that z_0 is a star for $R_1 \cap R_2$ and $R_1 \cup R_2$.

2. In various places in the course we shall use triangles, so we take the opportunity of getting our terminology precise. A *triangular contour* is a polygonal line of the form $[a, b, c, a]$. If a, b, c do not lie in a straight line the contour is simple-closed, and obviously has an inside which we call an *open triangle*.

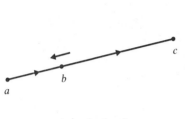 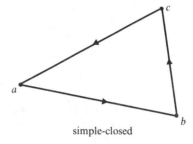

not simple-closed   simple-closed

Fig. 10

Prove that any open triangle is convex.

3. (a) Prove that any open sector $\{z : 0 < |z| < r, 0 < \text{Arg } z < \alpha\}$, where $\alpha \leqslant \pi$, is convex.
    (b) Prove that the punctured disc $D = \{z : 0 < |z| < 1\}$ is not a star region.

4. Let α be a point in $\mathbf{C}$, and θ some real number. The *ray from α with angle θ* is the set $L = \{\alpha + re^{i\theta} : r \geqslant 0\}$: see Fig. 11. Let R be the complement of L. Show that R is a star region. (Such a region R is called a cut plane. Cut planes crop up in connection with the logarithm functions: see *Unit 3, Differentiation.*)

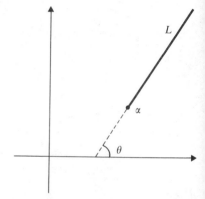

Fig. 11

5. Use previous problems to prove that each of the following is a star region. (We shall use the result of (c) in Section 5.5.)

(a)

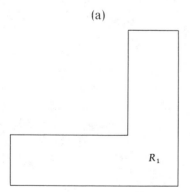

R_1

Fig. 12

(b)

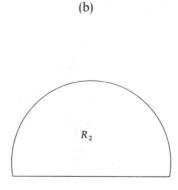

R_2

Fig. 13

(c)

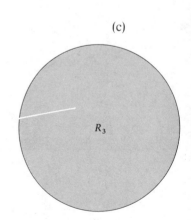

R_3

Fig. 14

6. Let R be a star region, a be a point not in R. Show that there is a ray from a not meeting R. (Hence, any star region is contained in a cut plane.)

Problem 7 deals with the Jordan Curve Theorem for a special type of arc: a polar arc. Before posing the problem we explain what a polar arc is. If you are pressed for time just read the following material, the problem and its solution.

We mentioned in *Unit 4* that the inside of any simple-closed arc is a region (this is part of the Jordan Curve Theorem). The regions that crop up in complex analysis very often arise in this way. There is a special kind of arc whose inside is always a star region, and these we discuss now. In fact, we shall actually prove the Jordan Curve Theorem for this particular kind of arc. This is perhaps a little surprising, as the general form of the Jordan Curve Theorem is excessively hard to prove.

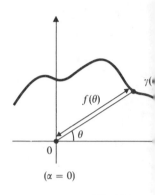

$(\alpha = 0)$

Fig. 15

We call an arc γ **polar about a point** α if there is a real function f, continuous and positive, with domain $[0, 2\pi]$ and $f(0) = f(2\pi)$, such that $\gamma(\theta) = f(\theta)e^{i\theta} + \alpha$, $\theta \in [0, 2\pi]$. This implies that γ is a simple-closed arc and that α does not lie on γ. It is also easy to see that $|f(\theta)e^{i\theta}| = |\gamma(\theta) - \alpha| = f(\theta)$. Fig. 15 illustrates the case $\alpha = 0$.

The standard example is the circle with centre α and radius r, given by the arc $\gamma(\theta) = re^{i\theta} + \alpha$: in this case $f(\theta) = r$. (If you have studied polar coordinate geometry you may know that ellipses, hyperbolas and parabolas can be given in similar form. In fact, an arc is polar about α precisely if it is a polar arc, relative to the pole α, in the sense of the Appendix to *Unit M231 10, Some Important Functions*.)

In the next problem you are asked to prove the following result.

The Jordan Curve Theorem for polar arcs

Let γ be an arc, polar about some point α. Let Γ be the path of γ. Then the complement of Γ is the union of two disjoint regions R_1 and R_2, where R_1 is a bounded region with star α, and R_2 is an unbounded region. Each of R_1 and R_2 has boundary Γ.

(Note: The best way to do this problem is to try each part in turn: if you cannot do one, look up the relevant part of the solution and then try the next.)

---

7. Let γ be an arc, polar about α. Let Γ be the path of γ. Let

$$R_1 = \{z : z = re^{i\theta} + \alpha \text{ for some } \theta \text{ and } r \text{ with}$$

$$0 \leqslant \theta \leqslant 2\pi \text{ and } 0 \leqslant r < |\gamma(\theta) - \alpha|\}.$$

(i) Show that R_1 is open. (Hint: Prove that every point of R_1 is an interior point. A figure would be helpful.)

(ii) Show that α is a star for R_1.

(iii) Show that R_1 is bounded.

Clearly R_1 above is the inside of γ. All we have to do now is define the outside and prove that it is a region.

(iv) Define the outside R_2. (Hint: What is left over?)

(v) Show that R_2 is open and unbounded.

(vi) Prove that R_2 is path-connected. (Hint: Can you use a curve like γ? Draw a picture.)

(vii) Why do R_1 and R_2 have boundary Γ?

Solutions

1. (a) Let $z_1, z_2 \in R_1 \cap R_2$. If $z \in [z_1, z_2]$ then $z \in R_1$ and $z \in R_2$, so that $z \in R_1 \cap R_2$.
   (b) See Fig. 16.

 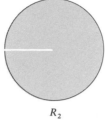 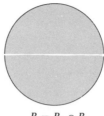

R_1 R_2 $R = R_1 \cap R_2$

Fig. 16

(c) Let $z \in R_1 \cap R_2$. Then $[z_0, z] \subseteq R_1$ and $[z_0, z] \subseteq R_2$, so that $[z_0, z] \subseteq R_1 \cap R_2$. Hence z_0 is a star for $R_1 \cap R_2$. If $z \in R_1 \cup R_2$ then $z \in R_1$ or $z \in R_2$, but in either case $[z_0, z] \subseteq R_1 \cup R_2$.

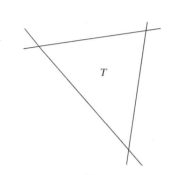

Fig. 17

2. Any open triangle T is the intersection of three open half-planes, and so is convex (Fig. 17).

3. (a) Any such sector is the intersection of two (or one) half-planes and a disc, and so convex.

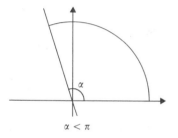 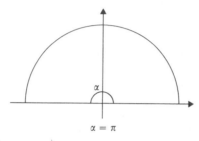 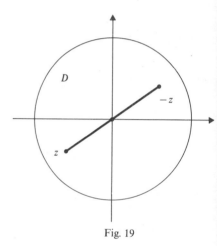

$\alpha < \pi$ $\alpha = \pi$

Fig. 18

Fig. 19

(b) Let z be a star for D. Consider $-z$, the reflection of z in the origin. Clearly $[-z, z]$ contains 0 and so does not lie wholly in D; but $-z \in D$ because $0 < |z| = |-z| < 1$ (Fig. 19).

4. By translation and rotation we can assume that L is the interval $(-\infty, 0]$ of $\mathbf{R}$.

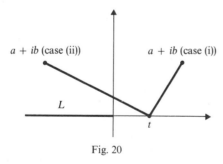

$a + ib$ (case (ii)) $a + ib$ (case (i))

L

t

Fig. 20

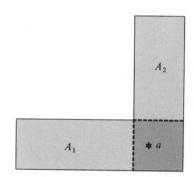

Fig. 21

Then any point t on the positive real axis is a star for R. For if $a + ib \in R$, then either (i) $a > 0$, and so $[t, a + ib]$ lies in $\{z : \operatorname{Re} z > 0\}$ and so in R, or (ii) $b \neq 0$, in which case $[t, a + ib]$ meets $\mathbf{R}$ only at t, and so lies in R.

5. (a) Any rectangle is the intersection of four half-planes and so convex. Now R_1 is the union of two rectangles A_1 and A_2 (Fig. 21). Let $a \in A_1 \cap A_2$. Then a is a star for $A_1 \cup A_2 = R_1$.

   (b) R_2 is the intersection of a disc and a half-plane, and so is convex.

   (c) Let L be the ray and D the open disc shown in Fig. 22. Then $R_3 = (\mathbf{C} - L) \cap D$. Any point a as shown in Fig. 22 is a star for the complement of L (by Problem 4), and for D, and hence for R_3.

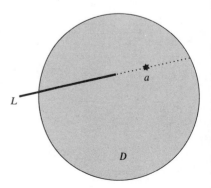

Fig. 22

6. Suppose that every ray from a contains a point of R. Let b be a star for R. Then the ray from a with direction opposite to $[a, b]$ contains a point $c \in R$.
Thus $[b, c] \subseteq R$ and so $a \in R$. This is a contradiction.

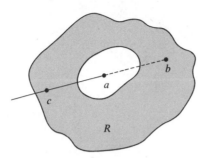

Fig. 23

7. (i) It is helpful to use $f(\theta) = |\gamma(\theta) - \alpha|$. Let $z_0 = r_0 \exp(i\theta_0) + \alpha$ lie in R_1. Then $r_0 < |\gamma(\theta_0) - \alpha| = f(\theta_0)$. Since f is continuous, there is $\delta > 0$ such that if $|\theta - \theta_0| < \delta$ then $|f(\theta) - f(\theta_0)| < \frac{1}{2}[f(\theta_0) - r_0]$ and so $f(\theta) > \frac{1}{2}[f(\theta_0) + r_0]$.

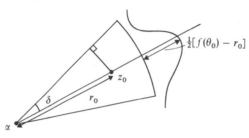

Fig. 24

Let $\varepsilon = \min(\frac{1}{2}[f(\theta_0) - r_0], r_0 \sin \delta)$. Thus the open disc $\{z : |z - z_0| < \varepsilon\}$ lies in R_1, for if $z = re^{i\theta} + \alpha$ is such that $|z - z_0| < \varepsilon$ then $|\theta - \theta_0| < \delta$ and so

$$f(\theta) > \frac{1}{2}[f(\theta_0) + r_0]$$
$$> r,$$

since $r \leqslant r_0 + \varepsilon \leqslant \frac{1}{2}[f(\theta_0) + r_0]$. (In order to deal with the case $\theta = 0$ it is best to assume that f has been extended to have domain $\mathbf{R}$ by using the definition $f(\theta + 2n\pi) = f(\theta)$.)

(ii) Let $z = re^{i\theta} + \alpha$ lie in R_1. If $z' \in [\alpha, z]$ then $z' = se^{i\theta} + \alpha$ for some $s < r$, and so $z' \in R_1$: thus $[\alpha, z] \subseteq R_1$ (Fig. 25).

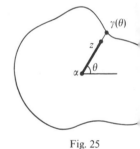

Fig. 25

(iii) Let K be an upper bound for f on $[0, 2\pi]$, by the Boundedness Theorem. Then $R_1 \subseteq \{z : |z - \alpha| < K\}$.

(iv) The outside is
$$R_2 = \{z : z = re^{i\theta} + \alpha, \text{ for some } \theta \text{ and } r \text{ with}$$
$$0 \leqslant \theta \leqslant 2\pi \text{ and } r > |\gamma(\theta) - \alpha|\}.$$

(v) Arguing as in (i), we see that R_2 is open. If $z \in R_2$ and $z \in [\alpha, z']$ then $|z' - \alpha| \geqslant |z - \alpha|$ and so $z' \in R_2$. Hence R_2 is unbounded.

(vi) Let $z_1 = r_1 \exp(i\theta_1) + \alpha$ and $z_2 = r_2 \exp(i\theta_2) + \alpha$ lie in R_2. (We can assume that $\theta_1 \leqslant \theta_2$.)
Let $c = r_1 - f(\theta_1)$: then $c > 0$.
Let $\gamma_1(\theta) = [f(\theta) + c]e^{i\theta} + \alpha, \theta_1 \leqslant \theta \leqslant \theta_2$.
Then γ_1 followed by the line segment $[\gamma_1(\theta_2), z_2]$ is a contour joining z_1 to z_2 (Fig. 26) lying entirely in R_2.

(vii) Note that $\gamma(\theta) = \lim_{r \to f(\theta)^-} (re^{i\theta} + \alpha)$: thus $\gamma(\theta)$ is a cluster point of R_1, and so a boundary point of R_1. Since $R_1 \cup \Gamma$ is closed (because it is the complement of the open set R_2), all boundary points of R_1 lie on Γ. Thus Γ is the boundary of R_1. Similarly for R_2.

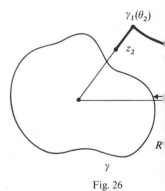

Fig. 26

62

5.3 CAUCHY'S THEOREM

In the last reading section we suggested that an obvious candidate for an "antiderivative" of a function f on a star region R was the function

$$F(z) = \int_{[z_0, z]} f, \qquad z \in R,$$

where z_0 is a star for R.

What we have to do now is to prove that F is analytic on R and that $F'(z) = f(z)$, or in other words, to show that

$$\lim_{h \to 0} \frac{F(z + h) - F(z)}{h} = f(z).$$

Now

$$\frac{F(z + h) - F(z)}{h} = \frac{1}{h}\left(\int_{[z_0, z+h]} f - \int_{[z_0, z]} f \right).$$

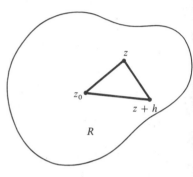

Fig. 27

If this were the proof of the Fundamental Theorem of Calculus in real analysis, we could replace the right-hand side above by $\dfrac{1}{h}\displaystyle\int_{[z, z+h]} f$. You should think about this for a moment: it is the crux of the matter. Let us look at a picture (Fig. 27) of the situation in order to see what this step would imply. There is an obvious closed contour—the triangular contour $\Delta = [z_0, z, z + h, z_0]$. Suppose we could prove that $\displaystyle\int_\Delta f = 0$, in other words that

$$\int_{[z_0, z]} f + \int_{[z, z+h]} f + \int_{[z+h, z_0]} f = 0.$$

Then by rearranging we would have

$$\int_{[z_0, z+h]} f - \int_{[z_0, z]} f = \int_{[z, z+h]} f,$$

which is just what we need in order to establish that

$$\frac{F(z + h) - F(z)}{h} = \frac{1}{h} \int_{[z, z+h]} f.$$

Since f is continuous at z, we may well be able to prove, by an ε, δ-argument and estimation of the integral, that the right-hand expression has limit $f(z)$ as h approaches 0. Thus it looks as if the key to our construction of a proposed antiderivative of f is a result which tells us that if the contour Δ is triangular then $\displaystyle\int_\Delta f = 0$. In fact, we have to be a little more careful with the hypotheses—here is the precise result.

Theorem 1 (The Cauchy–Goursat Theorem)*

Let f be analytic on a region R, and Δ be a triangular contour in R whose inside is also contained in R. Then $\displaystyle\int_\Delta f = 0$.

---

* *Augustin Louis Cauchy* (1789–1857), the father of complex analysis, was a profound and prolific writer who was known for his work both in real analysis, and in complex analysis. You have met the Cauchy–Riemann equations in *Unit 3*: in this unit and in *Unit 9* you will meet Cauchy's Theorem and Cauchy's Formulas.

Edouard Goursat (1858–1936) was instrumental in modifying Cauchy's Theorem.

Before considering the proof we make some further remarks about triangles. We define a *closed triangle* to be a triangular contour together with its inside if it is simple-closed—if the contour is not simple-closed we define the contour itself to be a *closed triangle*.

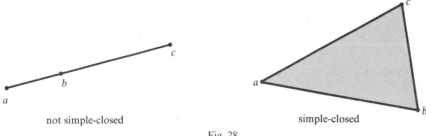

not simple-closed   simple-closed

Fig. 28

Where the context makes it clear we shall use the term *triangle* to mean the appropriate one of: triangular contour, open triangle, closed triangle.

Outline of Proof

We wish to show that the complex number $\int_\Delta f$ is 0, and we will achieve this by showing that $\left| \int_\Delta f \right| \le a_n$ where the sequence $\{a_n\}$ has limit zero. In fact, the elements a_n are of the form $\left| 4^{n-1} \int_{\Delta_n} f \right|$ where $\{\Delta_n\}$ is a well chosen sequence of triangles which gets smaller and smaller, each one lying in the previous one. A key point in the proof is to find a point lying inside or on all of the triangles Δ_n. This more or less follows from the Nested Rectangles Theorem. However, we shall prove a more useful and general lemma on intersection of closed sets.

Lemma

Let S be a bounded closed set, and $S_1, S_2, \ldots$ a sequence of nonempty closed subsets of S such that $S_n \supseteq S_{n+1}$, $n \in \mathbf{N}$. Then there is a point z such that $z \in S_n$ for all $n \in \mathbf{N}$: in other words, the intersection of the S_n is nonempty.

Proof

We use a compactness argument to obtain a contradiction. Suppose that every point z of S lies outside at least one of the S_n, that is suppose that for each $z \in S$ there is $n(z) \in \mathbf{N}$ such that $z \notin S_{n(z)}$. Since $S_{n(z)}$ is closed there is an open disc D_z with centre z which does not meet $S_{n(z)}$, that is $D_z \cap S_{n(z)} = \varnothing$. Since S is compact there is a finite number $z_1, \ldots, z_k$ of elements of S such that $S \subseteq D_{z_1} \cup \cdots \cup D_{z_k}$. Let $N = \max(n(z_1), \ldots, n(z_k))$. If $z \in S_N$ then $z \in D_{z_j}$ for some j (because $S_N \subseteq S$) and so $z \notin S_{n(z_j)}$, so that $z \notin S_N$ (because $N \ge n(z_j)$). Hence S_N is empty: but this is a contradiction since S_N is given to be nonempty. Thus for some $z \in S$ there is no $n \in \mathbf{N}$ such that $z \notin S_n$, that is, z lies in each S_n. ∎

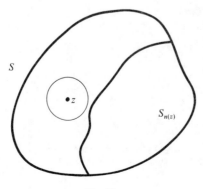

Fig. 29

There are two other points which crop up frequently in this and later units: they both occur in the next proof.

1. Recall from *Unit 4* that if Γ and $\Gamma_1, \ldots, \Gamma_n$ are contours such that $\Gamma = \Gamma_1 + \cdots + \Gamma_n$ and f is continuous on Γ, then

$$\int_\Gamma f = \int_{\Gamma_1} f + \cdots + \int_{\Gamma_n} f.$$

We often need an extension of this, when certain segments of $\Gamma_1, \ldots, \Gamma_n$ add together to give Γ and the rest cancel in pairs—in that case we still have

$$\int_\Gamma f = \int_{\Gamma_1} f + \cdots + \int_{\Gamma_n} f. \text{ Fig. 30 shows a simple example.}$$

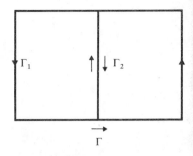

$\Gamma_1, \Gamma_2, \Gamma$ closed contours

Fig. 30

2. A standard method of proving that a complex number I is zero is to show that $|I| \leqslant \varepsilon$ for all positive ε. Note that this condition *does* actually imply that $I = 0$: for if $I \neq 0$ then $|I| > 0$ and by taking $\varepsilon = \frac{1}{2}|I|$ we obtain $|I| \leqslant \frac{1}{2}|I|$, which is a contradiction. Note also that I is often a contour integral, in other words, $I = \int_\Gamma f$ for suitable f and Γ. Do not forget: $\int_\Gamma f$ is a complex number.

Proof of Theorem 1

Step 1 We construct a suitable sequence $\{\Delta_n\}$ of triangles such that $\left| \int_\Delta f \right| \leqslant 4^{n-1} \left| \int_{\Delta_n} f \right|$ for all $n \in \mathbf{N}$.

Let $\Delta_1 = \Delta$. Clearly $\left| \int_\Delta f \right| \leqslant 4^0 \left| \int_{\Delta_1} f \right|$. Now suppose that Δ_n has just been constructed, and $\left| \int_\Delta f \right| \leqslant 4^{n-1} \left| \int_{\Delta_n} f \right|$.

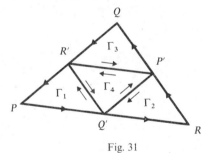

Fig. 31

By joining the midpoints of the sides of Δ_n we obtain four triangles Γ_1, Γ_2, Γ_3, Γ_4, as shown in Fig. 31. By following the contour $PQ'R'Q'RP'Q'P'QR'P'R'P$ it is clear that $\Delta_n = \Gamma_1 + \Gamma_2 + \Gamma_3 + \Gamma_4$ (in an obvious extension of the usual terminology), and so

$$\int_{\Delta_n} f = \int_{\Gamma_1} f + \int_{\Gamma_2} f + \int_{\Gamma_3} f + \int_{\Gamma_4} f.$$

Thus

$$\left| \int_{\Delta_n} f \right| \leqslant \left| \int_{\Gamma_1} f \right| + \left| \int_{\Gamma_2} f \right| + \left| \int_{\Gamma_3} f \right| + \left| \int_{\Gamma_4} f \right|.$$

Let k be such that $\left| \int_{\Gamma_k} f \right|$ is the largest of $\left| \int_{\Gamma_1} f \right|, \ldots, \left| \int_{\Gamma_4} f \right|$. Then $\left| \int_{\Delta_n} f \right| \leqslant 4 \left| \int_{\Gamma_k} f \right|$. Now let $\Delta_{n+1} = \Gamma_k$: then

$$\left| \int_\Delta f \right| \leqslant 4^{n-1} \left| \int_{\Delta_n} f \right| \leqslant 4^n \left| \int_{\Delta_{n+1}} f \right|.$$

(We have now constructed a sequence of triangles Δ_n such that

$$\left| \int_\Delta f \right| \leqslant 4^{n-1} \left| \int_{\Delta_n} f \right|$$

for all n; it remains to show that $\lim\limits_{n \to \infty} 4^{n-1} \left| \int_{\Delta_n} f \right| = 0$.

The essence of the next part of the proof is that we estimate $\left| \int_{\Delta_n} f \right|$ by finding the length of Δ_n and an upper bound for $|f|$ on Δ_n: we then use the Estimation Theorem of *Unit 4*.)

Step 2   We obtain the length of Δ_n.

Look back to the construction of Δ_{n+1}. A little geometry establishes that the length of each side of Δ_{n+1} is half the length of a side of Δ_n. (Clearly $PQ' = \frac{1}{2}PR$, and so on; the less obvious fact is that $P'Q' = \frac{1}{2}QP$, and so on—this is easily proved using similar triangles.) Thus if s_n is the length of the contour Δ_n (which is just the sum of the lengths of the sides), then $s_{n+1} = \frac{1}{2}s_n$. Let s be the length of Δ; then, by induction, $s_n = (\frac{1}{2})^{n-1}s$.

(We now require an upper bound for $|f|$ on Δ_n, but if you consider our estimate $\left| \int_\Delta f \right| \leqslant 4^{n-1} \left| \int_{\Delta_n} f \right|$, and the fact that the length of Δ_n is $(\frac{1}{2})^{n-1}s$, you will realize that our upper bound will need to be very small, better even than $(\frac{1}{2})^{n-1}$, for the method to work. First we show that for sufficiently large n the triangles Δ_n are very near to a point ζ of R, and then, that, for z on Δ_n, $f(z)$ is very nearly $f(\zeta)$.)

Step 3   There is a point ζ belonging to all the triangles Δ_n.

Let T_n be the closed triangle corresponding to the triangular contour Δ_n. Then T_n is a nonempty closed set, and, furthermore, $T_n \supseteq T_{n+1}$ for all n. Since T_1 is closed and bounded, by the lemma there is a point ζ such that $\zeta \in T_n$ for all n.

Step 4   We show that for large n, T_n is close to ζ.

If z is any point inside or on the triangle Δ_n, then the distance from z to ζ is certainly less than the length s_n of the perimeter of T_n. Hence, for any $z \in T_n$,

$$|\zeta - z| \leqslant s_n = (\tfrac{1}{2})^{n-1}s.$$

Step 5   We use the fact that f is differentiable at ζ to approximate f on Δ_n.

Since f is differentiable at ζ, we know that

$$\lim_{z \to \zeta} \left(\frac{f(z) - f(\zeta)}{z - \zeta} - f'(\zeta) \right) = 0,$$

and it follows that

$$f(z) = f(\zeta) + f'(\zeta)(z - \zeta) + (z - \zeta)r(z), \qquad\qquad (*)$$

where

$$r(z) = \begin{cases} \dfrac{f(z) - f(\zeta)}{z - \zeta} - f'(\zeta), & z \neq \zeta \\[2mm] 0, & z = \zeta, \end{cases}$$

and $\lim\limits_{z \to \zeta} r(z) = 0$.

(You might expect that we could now use (*) to give an upper estimate for $|f|$ on T_n and thus estimate $\left| \int_{\Delta_n} f \right|$. We could certainly do this, but it would not give a sufficiently small estimate. The next step is to split the expression on the right of (*) into two parts both of which contribute something very small to $\left| \int_{\Delta_n} f \right|$, but for different reasons. In fact one of the parts contributes nothing!)

Step 6 We obtain an estimate for $\left| \int_{\Delta_n} f \right|$.

We write

$$f(z) = \boxed{f(\zeta) + f'(\zeta)(z - \zeta)} + (z - \zeta)r(z),$$

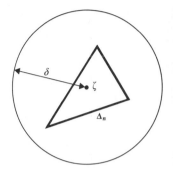

Fig. 32

and notice that the boxed expression on the right is the derivative at z of $z \longrightarrow zf(\zeta) + f'(\zeta)\dfrac{(z - \zeta)^2}{2}$, and therefore

$$\int_{\Delta_n} [f(\zeta) + f'(\zeta)(z - \zeta)]\, dz = 0.$$

On the other hand, f is differentiable at ζ and we therefore know that, given $\varepsilon > 0$, there is a $\delta > 0$ such that

$$|r(z)| < \varepsilon \quad \text{if } |z - \zeta| < \delta.$$

If we now ensure that Δ_n lies inside the disc $|z - \zeta| < \delta$, by choosing $(\tfrac{1}{2})^{n-1}s < \delta$ then, by the Estimation Theorem,

$$\left| \int_{\Delta_n} f \right| = \left| \int_{\Delta_n} (z - \zeta)r(z)\, dz \right|$$

$$\leqslant (\text{the length of } \Delta_n)$$

$$\times (\text{an upper estimate for } z \longrightarrow |z - \zeta|\,|r(z)| \text{ on } \Delta_n)$$

$$\leqslant (\tfrac{1}{2})^{n-1}s \cdot (\tfrac{1}{2})^{n-1}s \cdot \varepsilon$$

$$= \frac{\varepsilon s^2}{4^{n-1}}.$$

Finally, we have

$$\left| \int_{\Delta} f \right| \leqslant 4^{n-1} \left| \int_{\Delta_n} f \right| \leqslant 4^{n-1} \cdot \frac{\varepsilon s^2}{4^{n-1}} = \varepsilon s^2$$

for all n sufficiently large. It follows that $\int_{\Delta} f = 0$, which completes the proof. ∎

We needed this theorem to fix up our construction of an antiderivative of a function analytic on a star region. We can now give the full argument, but first we state the definition and introduce some notation which is useful in conjunction with estimating integrals.

Definition

> Let f be defined on a region R. A function F such that $F' = f$ on R is called an **antiderivative of f on R**.

Notation

We shall write $\int_{\gamma} f(z)|dz|$ for $\int_a^b f(\gamma(t))|\gamma'(t)|\, dt$ where γ is a smooth arc with domain $[a, b]$. If γ has path Γ then $\int_{\Gamma} f(z)|dz|$ is defined to be $\int_{\gamma} f(z)|dz|$, and if $\Gamma = (\Gamma_1, \ldots, \Gamma_n)$ is a contour then $\int_{\Gamma} f(z)|dz|$ is defined to be $\sum_{i=1}^{n} \int_{\Gamma_i} f(z)|dz|$.

The notation seems reasonable when we consider the substitution $z = \gamma(t)$, $dz = \gamma'(t)dt$; for then naturally $|dz| = |\gamma'(t)|dt$. In this notation the key part of the Estimation Theorem of *Unit 4* is:

$$\left| \int_\Gamma f(z)\, dz \right| \leqslant \int_\Gamma |f(z)| \cdot |dz|.$$

Theorem 2 (The Antiderivative Theorem)

Let f be analytic on a star region R. Then f has an antiderivative on R, that is, there is a function F analytic on R such that

$$F'(z) = f(z), \qquad z \in R.$$

Proof

Let z_0 be a star for R. Define F by $F(z) = \displaystyle\int_{[z_0,z]} f$, $z \in R$. We shall show that for $z \in R$, $F'(z) = f(z)$. Fix a particular $z \in R$. Since R is open, there is a $\delta > 0$ such that the disc $\{\zeta : |\zeta - z| < \delta\}$ lies in R. Let $0 < |h| < \delta$. Then

$$\frac{F(z + h) - F(z)}{h} = \frac{1}{h}\left(\int_{[z_0, z+h]} f - \int_{[z_0, z]} f \right).$$

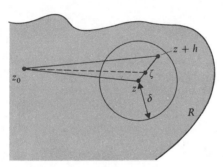

Fig. 33

Consider the triangle $\Delta = [z_0, z, z + h, z_0]$. Each point on or inside this triangle lies on a line segment $[z_0, \zeta]$ for some $\zeta \in [z, z + h]$. But all such ζ lie in R, and so, because z_0 is a star, $[z_0, \zeta] \subseteq R$. Thus Δ and its inside lie in R and so the Cauchy–Goursat Theorem implies that $\displaystyle\int_\Delta f = 0$. Hence

$$\frac{F(z + h) - F(z)}{h} = \frac{1}{h} \int_{[z, z+h]} f(\zeta)\, d\zeta,$$

and so

$$\frac{F(z + h) - F(z)}{h} - f(z) = \frac{1}{h} \int_{[z, z+h]} (f(\zeta) - f(z))\, d\zeta.$$

Let $\varepsilon > 0$. Since f is continuous at z, there is δ_1 such that $0 < \delta_1 \leqslant \delta$ and $|f(\zeta) - f(z)| < \varepsilon$ if $|\zeta - z| < \delta_1$. If $0 < |h| < \delta_1$ then

$$\left| \frac{1}{h} \int_{[z, z+h]} (f(\zeta) - f(z)) d\zeta \right| \leqslant \frac{1}{|h|} \int_{[z, z+h]} \varepsilon |d\zeta|$$

$$= \frac{1}{|h|} \varepsilon |h| = \varepsilon.$$

Hence

$$\lim_{h \to 0} \left(\frac{F(z + h) - F(z)}{h} - f(z) \right) = 0.$$

In other words, F is differentiable at z and $F'(z) = f(z)$. ∎

It is worth pointing out that in complex analysis theorems where we have to prove that a function ψ is the derivative of a function ϕ we nearly always express $\dfrac{\phi(z+h)-\phi(z)}{h} - \psi(z)$ as an integral $I(h)$, and then estimate $I(h)$ (that is, bound $|I(h)|$ above) by an expression $J(h)$ where $\lim\limits_{h\to0} J(h) = 0$. Further examples will occur in the theorems of the next reading section.

We shall extend Theorem 2 to a wider class of regions in *Unit 9*. However, as we have already pointed out, there are *some* regions R on which some functions f analytic on R do *not* have antiderivatives—for example $R = \{z \in \mathbf{C}: z \neq 0\}$ and $f(z) = 1/z$. On the other hand, the function $g(z) = 1/z^2$ *does* have an anti-derivative on R, namely $G(z) = -1/z$.

The Antiderivative Theorem would seem to be mainly a theoretical tool: the antiderivatives it provides are not much use for actually evaluating a contour integral $\displaystyle\int_\Gamma f$. However, there is a consequence of it which leads to a whole host of theorems and practical ways of evaluating integrals. In fact, this consequence is much more important than the theorem, which is why it has given its name to the unit. Here it is.

Theorem 3 (Cauchy's Theorem for Star Regions)

Let f be any function analytic on a star region R, and Γ be any closed contour in R. Then $\displaystyle\int_\Gamma f = 0$.

Proof

By the Antiderivative Theorem there is an antiderivative F of f on R. By the Fundamental Theorem for contour integrals (Theorem 8 of *Unit 4*), $\displaystyle\int_\Gamma f = 0$, since Γ is closed. ∎

This result is the central theorem of complex analysis. To say any more at this stage would be an anticlimax.

Summary

In this section we have proved Cauchy's Theorem for star regions. We began by proving the Cauchy–Goursat Theorem, using a nested triangles argument. This theorem ensured that a certain construction of a supposed antiderivative for a function analytic on a star region did in fact work, and from this last result (the Antiderivative Theorem) it was easy to deduce Cauchy's Theorem for star regions.

Self-Assessment Questions

1. Let $D = \{z:|z| < 1\}$. Which of the following functions f and contours Γ shown in the figures satisfy the hypotheses of Cauchy's Theorem in D? In each case the circle is $|z| = 1$.

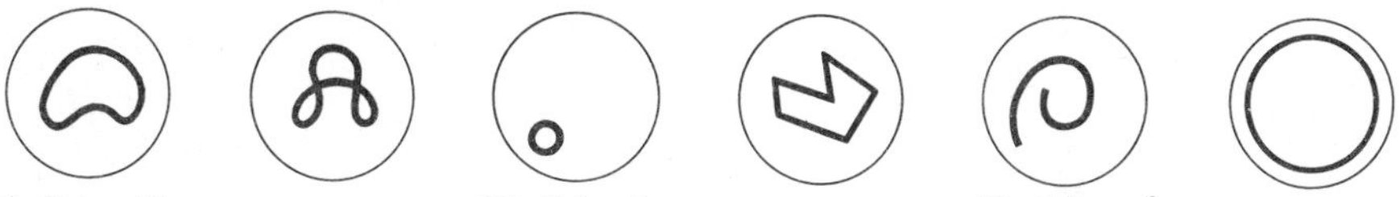

(A) $f(z) = 1/z$ (C) $f(z) = \operatorname{Log} z$ (E) $f(z) = z^2$

(B) $f(z) = e^z$ (D) $f(z) = 1/(z-3)$ (F) $f(z) = |e^z|$

Fig. 34

69

2. What are the main steps in the proof of the Cauchy–Goursat Theorem for a triangle Δ in a region R?

Solutions

1. (A) No: f is not analytic on D.
   (B) Yes: even though Γ is not simple.
   (C) No: Log is not analytic on all of D.
   (D) Yes: f is analytic just on D (outside does not matter).
   (E) No: Γ is not closed.
   (F) No: f is not analytic on D.

2. (1) Construct a sequence of triangles $\{\Delta_n\}$ with $\Delta_1 = \Delta$ such that
$$\left| \int_{\Delta_n} f \right| \leqslant 4 \left| \int_{\Delta_{n+1}} f \right| \text{ for all } n.$$
   (2) Obtain the length of Δ_n.
   (3) Show that there is a point ζ belonging to all the triangles Δ_n.
   (4) Show that $|\zeta - z| \leqslant (\tfrac{1}{2})^{n-1}s$ where s is the length of Δ.
   (5) Show that on Δ_n, $f(z)$ may be expressed as
$$f(\zeta) + f'(\zeta)(z - \zeta) + (z - \zeta)r(z),$$
   where $\lim_{z \to \zeta} r(z) = 0$.

   (6) Obtain an estimate for $\left| \int_{\Delta_n} f \right|$ by using the Estimation Theorem. In fact, given $\varepsilon > 0$,
$$\left| \int_{\Delta_n} f \right| \leqslant \frac{\varepsilon s^2}{4^{n-1}}.$$
   Hence
$$\left| \int_{\Delta} f \right| = 0.$$

5.4 PROBLEMS

Most of the applications of Cauchy's Theorem in complex analysis derive from various *theoretical* consequences of Cauchy's Theorem which we shall not discuss until the next section. However, Cauchy's Theorem can be used directly to evaluate certain real integrals. For if $\int_\gamma f(z)\,dz = 0$, then

$$0 = \int_a^b f(\gamma(t))\gamma'(t)\,dt \qquad \text{(where } \gamma \text{ has domain } [a, b])$$

$$= I + iJ \quad \text{for certain real integrals } I \text{ and } J,$$

and so $I = J = 0$.

We give an example of this in Problem 1. Further examples, dealing with the evaluation of so-called *improper* integrals $\left(\text{such as } \int_0^\infty \frac{\sin x}{x}\,dx\right)$, will be discussed in *Unit 10, The Calculus of Residues*.

---

1. (a) Let C be any circle and b a point outside C. Show that $\mathrm{Wnd}\,(C, b)$, the winding number of C about b, is zero.

    (b) Now let C be the circle $|z| = a$ and b a real number such that $b > a$. By considering $\int_C \frac{dz}{z - b}$, evaluate two real integrals.

2. Let f be analytic on a region R (not necessarily a star region) and C be a rectangle in R whose inside is also contained in R. Adapt the Cauchy–Goursat proof to show that $\int_C f = 0$. (Decompose the rectangle into four similar ones. You need only write down the main steps of the argument.)

3. Use the Cauchy–Goursat Theorem to give a short proof of the result of Problem 2 by splitting C into two triangles. (This method of *triangulation* will be very important in *Unit 9*.)

4. Let R be a star region, f be analytic on R and Γ_1, Γ_2 be two contours in R with the same initial and final points. Show that $\int_{\Gamma_1} f = \int_{\Gamma_2} f$.

Solutions

1. (a) Let C be the circle $|z - \alpha| = r$. Let D be the open disc $\{z : |z - \alpha| < |b - \alpha|\}$. Then D is a star region containing the circle C, and the function $z \longrightarrow \dfrac{1}{z - b}$ is analytic on D. Hence by Cauchy's Theorem,

$$\int_C \frac{dz}{z - b} = 0.$$

Thus

$$\mathrm{Wnd}(C, b) = \frac{1}{2\pi i} \int_C \frac{dz}{z - b} = 0.$$

(b) Since b lies outside C, $\int_C \dfrac{dz}{z - b} = 0$ by part (a).

Now C is the path of the arc $\gamma(\theta) = ae^{i\theta}$, $\theta \in [0, 2\pi]$, and so

$$0 = \int_0^{2\pi} \frac{iae^{i\theta}}{ae^{i\theta} - b}\, d\theta$$

$$= ia \int_0^{2\pi} \frac{1}{a - b(\cos\theta - i\sin\theta)}\, d\theta$$

$$= ia \int_0^{2\pi} \frac{a - b\cos\theta - ib\sin\theta}{(a - b\cos\theta)^2 + (b\sin\theta)^2}\, d\theta$$

$$= \int_0^{2\pi} \frac{i(a^2 - ab\cos\theta) + ab\sin\theta}{a^2 - 2ab\cos\theta + b^2}\, d\theta.$$

Hence

$$\int_0^{2\pi} \frac{a^2 - ab\cos\theta}{a^2 - 2ab\cos\theta + b^2}\, d\theta = 0$$

and

$$\int_0^{2\pi} \frac{ab\sin\theta}{a^2 - 2ab\cos\theta + b^2}\, d\theta = 0.$$

2. Here is a fairly full solution. We construct a sequence $C_1, \ldots$ of rectangles as follows. First let $C_1 = C$. Now suppose that C_n has been constructed.

By joining the midpoints of the sides of C_n we obtain four rectangles T_1, T_2, T_3, T_4 with positive orientation, as shown in Fig. 35. Choose k such that $\left| \int_{T_k} f \right|$ is largest of

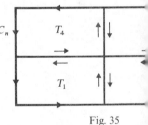

Fig. 35

$$\left| \int_{T_1} f \right|, \ldots, \left| \int_{T_4} f \right|,$$ and let $C_{n+1} = T_k$. Clearly $\int_{C_n} f = \sum_{j=1}^{4} \int_{T_j} f$, and so

$$\left| \int_{C_n} f \right| \leqslant 4 \left| \int_{C_{n+1}} f \right|.$$

By mathematical induction $\left| \int_C f \right| \leqslant 4^{n-1} \left| \int_{C_n} f \right|$. By the lemma for Theorem 1, there is a point ζ inside or on each C_n. Since f is differentiable at ζ,

$$f(z) = f(\zeta) + (z - \zeta)f'(\zeta) + (z - \zeta)r(z)$$

where r is continuous at ζ and $r(\zeta) = 0$. We now estimate $\left| \int_{C_n} f \right|$. Since $z \longrightarrow f(\zeta) + f'(\zeta)(z - \zeta)$ is a derivative, we have

$$\left| \int_{C_n} f \right| = \left| \int_{C_n} (z - \zeta)r(z)\, dz \right|.$$

Let $\varepsilon > 0$. Then, for sufficiently large n, $\left| \int_{C_n} f \right| \leqslant \varepsilon s_n^2$, by the Estimation Theorem, where s_n is the length of C_n. Clearly $s_n = \left(\tfrac{1}{2}\right)^{n-1} s$, where s is the length of C, and so

$$\left| \int_C f \right| \leqslant 4^{n-1} \left| \int_{C_n} f \right| \leqslant \varepsilon s^2.$$

It follows that $\int_C f = 0.$

3.

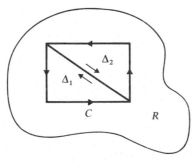

Fig. 36

Clearly the triangles Δ_1 and Δ_2 lie in R, and their insides lie inside C and so in R. Hence by the Cauchy–Goursat Theorem, $\int_{\Delta_1} f = \int_{\Delta_2} f = 0$. Since $\int_C f = \int_{\Delta_1} f + \int_{\Delta_2} f$, we have

$$\int_C f = 0.$$

In fact this method can be applied to any polygon C, however complicated—but "clearly" above becomes steadily less clear. Thus at this stage of the course we prefer to add the extra assumption that R is a star region, and deduce $\int_C f = 0$ by Cauchy's Theorem. However, in *Unit 9* the technique foreshadowed here will be discussed in some detail.

4. Let $\Gamma = \Gamma_1 + (-\Gamma_2)$. Then Γ is a closed contour in R and so $\int_\Gamma f = 0$ by Cauchy's Theorem. But

$$\int_\Gamma f = \int_{\Gamma_1 + (-\Gamma_2)} f = \int_{\Gamma_1} f - \int_{\Gamma_2} f,$$

and so $\int_{\Gamma_1} f = \int_{\Gamma_2} f$. (We sometimes say that in a star region, contour integrals of analytic functions are *path-independent*, since the value depends only on the end-points, not the particular path, of a contour.)

5.5 CAUCHY'S FORMULAS

One of the remarkable properties of functions analytic on a region, that you will see unfolded in this unit, and *Unit 6, Taylor Series, and Unit 9, Cauchy's Theorem II*, is that they are uniquely determined by their values at rather "few" points. The grandfather of all such results, and of many others in complex analysis, is Cauchy's Formula. We shall state it precisely and then discuss it.

Theorem 4 (Cauchy's Formula)

Let C be the circle $|z - \alpha| = r$, with inside D. Let f be analytic on some region R containing $D \cup C$. Then for any z in D,

$$f(z) = \frac{1}{2\pi i}\int_C \frac{f(\zeta)}{\zeta - z}d\zeta.$$

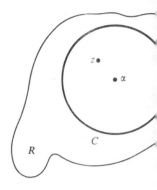

Fig. 37

In other words, f is uniquely determined in the open disc D by its values on the boundary C of D, and we can even give a nice formula for the values of f in D.

Later on in this unit and in *Unit 9* we shall discuss generalizations of Cauchy's Formula where the circle C is replaced by other contours. But the above form is sufficient for most *theoretical* purposes, although for *practical* calculation of integrals it is useful to have these generalized forms (some of which are discussed in the next problems section).

The proof of Cauchy's Formula uses estimation of an integral, and a slight extension of Cauchy's Theorem, which is of some interest in its own right.

Proof of Theorem 4

Fix $z \in D$. Let C' be a small circle centre z, of radius ρ, say. How small ρ is we shall decide later, but to begin with let $\rho < r - |z - \alpha|$, so that C' lies in D. Suppose that we could prove that

$$\int_C \frac{f(\zeta)}{\zeta - z}d\zeta = \int_{C'} \frac{f(\zeta)}{\zeta - z}d\zeta. \tag{1}$$

(In fact we *shall* prove it by means of a lemma after this proof.)

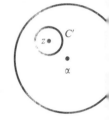

Fig. 38

Then we can proceed as follows. We know from Problem 6 of Section 4.6 of *Unit 4* that $2\pi i$ occurs as the value of a certain integral, namely

$$\int_{C'} \frac{1}{\zeta - z}d\zeta = 2\pi i. \tag{2}$$

Let $I = \int_C \frac{f(\zeta)}{\zeta - z}d\zeta - 2\pi i f(z)$. We wish to show that $I = 0$. By using (1) and (2),

$$I = \int_{C'} \frac{f(\zeta)}{\zeta - z}d\zeta - f(z)\int_{C'} \frac{1}{\zeta - z}d\zeta$$

$$= \int_{C'} \frac{f(\zeta) - f(z)}{\zeta - z}d\zeta.$$

We have control over the radius ρ of C', so let us try to make the last integral arbitrarily small by suitably varying ρ. Now

$$\left|\int_{C'} \frac{f(\zeta) - f(z)}{\zeta - z}d\zeta\right| \leqslant \int_{C'} \frac{|f(\zeta) - f(z)|}{|\zeta - z|}|d\zeta|$$

$$= \frac{1}{\rho}\int_{C'} |f(\zeta) - f(z)| \cdot |d\zeta|,$$

since $|\zeta - z| = \rho$ for $\zeta \in C'$.

74

Let $\varepsilon > 0$. Because f is continuous at z, there is $\delta > 0$ such that $|f(\zeta) - f(z)| < \varepsilon$ whenever $|\zeta - z| < \delta$. So, if we choose $\rho < \delta$, we can simplify our estimate even more, since

$$\frac{1}{\rho} \int_{C'} |f(\zeta) - f(z)| \cdot |d\zeta| \leqslant \frac{1}{\rho} \varepsilon \cdot 2\pi\rho = 2\pi\varepsilon.$$

Putting all these calculations together and remembering the definition of I, we have $|I| \leqslant 2\pi\varepsilon$. But this holds for *all* $\varepsilon > 0$. Hence $I = 0$, that is

$$\int_C \frac{f(\zeta)}{\zeta - z} d\zeta = 2\pi i f(z). \quad \blacksquare$$

Thus we have proved this theorem *if* we can show that

$$\int_C \frac{f(\zeta)}{\zeta - z} d\zeta = \int_{C'} \frac{f(\zeta)}{\zeta - z} d\zeta.$$

We shall prove this fact by establishing a rather more general lemma.

Lemma

Let C be the circle $\{\zeta : |\zeta - \alpha| = r\}$ and C' be the circle $\{\zeta : |\zeta - z| = \rho\}$, where $|z - \alpha| + \rho < r$ (so that C' lies inside C). Let g be analytic on some region containing C and its inside, except possibly for z. Then

$$\int_C g(\zeta) \, d\zeta = \int_{C'} g(\zeta) \, d\zeta.$$

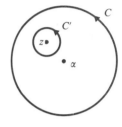

Fig. 39

If you are interested in generalities, you can view this lemma in two ways. The first is as a kind of Cauchy's Theorem for the region between C' and C. The second is as a "deformation theorem": the circle C can be continuously deformed (through the region of analyticity of g) onto C', and this does not change the value of the integral. But now, the proof.

Proof of Lemma

We have to show that $\int_C g(\zeta) \, d\zeta - \int_{C'} g(\zeta) \, d\zeta = 0$, or in a suggestive notation, that $\left(\int_C - \int_{C'} \right) g(\zeta) \, d\zeta = 0$. This integral looks as though it is trying to be

$$\int_{C+(-C')} g(\zeta) \, d\zeta = 0,$$

but unfortunately $C + (-C')$ is not defined because C and $-C'$ do not meet. However, can we define something *equivalent* to this, by adding extra lines, as we did in the Cauchy–Goursat Theorem? We can: join C and C' by lines as shown in Fig. 40 and let Γ_1 and Γ_2 be the two contours shown there.

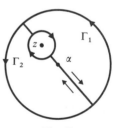

Fig. 40

By noting that integrals along the line segments cancel, we see that

$$\left(\int_C - \int_{C'} \right) g(\zeta) \, d\zeta = \left(\int_{\Gamma_1} + \int_{\Gamma_2} \right) g(\zeta) \, d\zeta.$$

We can now show that

$$\int_{\Gamma_1} g(\zeta) \, d\zeta = 0 \quad \text{and} \quad \int_{\Gamma_2} g(\zeta) \, d\zeta = 0,$$

provided we can find star regions to put Γ_1 and Γ_2 in. Certainly the insides of Γ_1 and Γ_2 look a bit like star regions, but unfortunately Cauchy's Theorem requires Γ_1 and Γ_2 actually to lie *in* star regions. Thus we have to "fatten up" the insides a little, but still keep g analytic on the fatter regions. This can be done by using the continuous function d_A introduced in *Unit 2*; recall that $d_A(\zeta)$ is the least distance from a point ζ to a (nonempty) set A. Let R be the region on which g is analytic, together with z; now let A be the complement of R. Assume first that A is nonempty. Then d_A has a minimum, δ say, on the closed disc $S = \{\zeta : |\zeta - \alpha| \leq r\}$, by the Corollary to Theorem 12 of *Unit 2*; if $\delta = 0$ then A would contain a point of S (since A is closed), so that $\delta > 0$. If A is empty let δ be any positive real number.

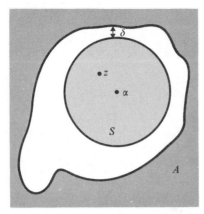

Fig. 41

In either case, the open disc $E = \{\zeta : |\zeta - \alpha| < r + \frac{1}{2}\delta\}$ is contained in R and contains C. Now we can find a star region contained in R but containing Γ_1, namely the set D_1 in Fig. 42. (It is star by Problem 5 in Section 5.2: notice the cut from z.) Certainly g is analytic on D_1 (because $z \notin D_1$ and $D_1 \subseteq E \subseteq R$).

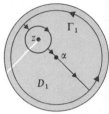

Fig. 42

Thus $\int_{\Gamma_1} g(\zeta)\, d\zeta = 0$. Similarly (by reflecting in the line through α and z), $\int_{\Gamma_2} g(\zeta)\, d\zeta = 0$. Hence

$$\left(\int_C - \int_{C'} \right) g(\zeta)\, d\zeta = 0. \quad \blacksquare$$

Now we can conclude the proof of Cauchy's Formula: we need just put $g(\zeta) = \dfrac{f(\zeta)}{\zeta - z}$ in the lemma to establish that

$$\int_C \frac{f(\zeta)}{\zeta - z}\, d\zeta = \int_{C'} \frac{f(\zeta)}{\zeta - z}\, d\zeta,$$

which was the statement (1) required in the proof of Theorem 4.

There are two ways of viewing Cauchy's Formula:

$$f(\alpha) = \frac{1}{2\pi i} \int_C \frac{f(z)}{z - \alpha}\, dz.$$

One is in the direction right to left, evaluating the integral $\int_C \dfrac{f(z)}{z - \alpha}\, dz$ as $2\pi i f(\alpha)$—this is a fairly practical use. The other is from left to right, expressing $f(\alpha)$ as an integral and using this representation to find properties of f—see Theorem 7 later in the section. This use is more theoretical.

First we discuss the *practical* use of Cauchy's Formula for evaluating various contour integrals around circles. For example, if C is the circle $\{z : |z| = 2\}$, and f is the function $z \longrightarrow e^z$, which is entire, then with $\alpha = 0$, we have

$$\frac{1}{2\pi i} \int_C \frac{e^z}{z}\, dz = e^0 = 1,$$

and with $\alpha = 1$, we have

$$\frac{1}{2\pi i} \int_C \frac{e^z}{z - 1}\, dz = e^1 = e.$$

Here is a less trivial example.

76

Example 1

Calculate $\int_C \dfrac{\cos z}{z^3 + z}\, dz$, where C is

(i)   the circle $\{z : |z| = 2\}$, and
(ii) the circle $\{z : |z + i| = \tfrac{1}{2}\}$.

Solution

Clearly, $z^3 + z = z(z^2 + 1) = z(z + i)(z - i)$. By the method of partial fractions (which is valid in complex analysis), we have

$$\frac{1}{z^3 + z} = \frac{1}{z} - \frac{\tfrac{1}{2}}{z + i} - \frac{\tfrac{1}{2}}{z - i},$$

and so

$$\int_C \frac{\cos z}{z^3 + z}\, dz = \int_C \frac{\cos z}{z}\, dz - \frac{1}{2}\int_C \frac{\cos z}{z + i}\, dz - \frac{1}{2}\int_C \frac{\cos z}{z - i}\, dz.$$

(i)   Because cos is entire and $0, i, -i$ lie inside C, we have, by Cauchy's Formula,

$$\int_C \frac{\cos z}{z^3 + z}\, dz = 2\pi i(\cos 0 - \tfrac{1}{2}\cos(-i) - \tfrac{1}{2}\cos i)$$

$$= 2\pi i(1 - \cosh 1).$$

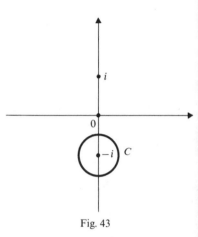

Fig. 43

(ii) Because $z \longrightarrow \dfrac{\cos z}{z}$ is analytic on the star region $R = \{z : \operatorname{Im} z < 0\}$ and C is contained in R (Fig. 43), we have $\int_C \dfrac{\cos z}{z}\, dz = 0$, by Cauchy's Theorem (Theorem 3).

Similarly, $\int_C \dfrac{\cos z}{z - i}\, dz = 0$, and so

$$\int_C \frac{\cos z}{z^3 + z}\, dz = -\frac{1}{2}\int_C \frac{\cos z}{z + i}\, dz$$

$$= -\tfrac{1}{2} \cdot 2\pi i \cos(-i), \quad \text{by Cauchy's Formula,}$$

$$= -\pi i \cosh 1.$$

By methods similar to those in the example we can deal with any integral of the form

$$\int_C \frac{f(z)}{p(z)}\, dz,$$

where C is a circle, f is a function analytic on a region R containing C and its inside, and p is a polynomial with *distinct* roots. But we cannot yet deal with an integral such as

$$\int_C \frac{e^z}{(z - 1)^2}\, dz.$$

However, there *is* an extension of Cauchy's Formula which covers such cases. This extension is a consequence of a theoretical result which is important for many reasons other than merely evaluating integrals.

Let us see how we can extend Cauchy's Formula:

$$f(z) = \frac{1}{2\pi i}\int_C \frac{f(\zeta)}{\zeta - z}\, d\zeta.$$

We can put this in more general terms. Let H be some complex-valued function of two complex variables, and $f(z) = \int_C H(\zeta, z)\, d\zeta$. It would now seem plausible

77

that

$$f'(z) = \frac{d}{dz}\int_C H(\zeta, z)\, d\zeta = \int_C \frac{\partial H}{\partial z}(\zeta, z)\, d\zeta,$$

where we write $\dfrac{\partial H}{\partial z}$ instead of $\dfrac{dH}{dz}$ because H is a function of two variables.

But of course there is no reason why, in general, integration and differentiation should change places so easily. Under certain assumptions on H they do, but if you try and prove this in a straightforward way you run into a brick wall. (We discuss this point again in *Unit 11, Analytic Functions*.) However, for very special H, such as the H in Cauchy's Formula, which is $H(\zeta, z) = \dfrac{1}{2\pi i}\dfrac{f(\zeta)}{\zeta - z}$,

we *can* prove it. Note that if $H(\zeta, z) = \dfrac{1}{2\pi i}\dfrac{f(\zeta)}{\zeta - z}$, then

$$\frac{\partial H}{\partial z}(\zeta, z) = \frac{1}{2\pi i} f(\zeta)\frac{-1}{(\zeta - z)^2}(-1) = \frac{1}{2\pi i}\frac{f(\zeta)}{(\zeta - z)^2};$$

thus we expect the formula for $f'(z)$ given in the following theorem.

Theorem 5 (Cauchy's Formula for the First Derivative)

Let f be analytic on some region containing the circle $C = \{z : |z - \alpha| = r\}$ and its inside D. Then for any z in D,

$$f'(z) = \frac{1}{2\pi i}\int_C \frac{f(\zeta)}{(\zeta - z)^2}\, d\zeta.$$

Proof

Let z be a point in D. Naturally enough we first write down $\dfrac{f(z + h) - f(z)}{h}$.

Using Cauchy's Formula, this equals

$$\frac{1}{h \cdot 2\pi i}\left(\int_C \frac{f(\zeta)}{\zeta - (z + h)}\, d\zeta - \int_C \frac{f(\zeta)}{\zeta - z}\, d\zeta\right)$$

$$= \frac{1}{2\pi i}\int_C f(\zeta)\frac{1}{h}\left(\frac{1}{\zeta - z - h} - \frac{1}{\zeta - z}\right) d\zeta$$

$$= \frac{1}{2\pi i}\int_C f(\zeta)\frac{1}{(\zeta - z - h)(\zeta - z)}\, d\zeta.$$

We must now show that the limit of this expression as h approaches 0 is the given expression for $f'(z)$; in other words, that their difference, $\phi(h)$, say, has limit 0. Now

$$\phi(h) = \frac{f(z + h) - f(z)}{h} - \frac{1}{2\pi i}\int_C \frac{f(\zeta)}{(\zeta - z)^2}\, d\zeta$$

$$= \frac{1}{2\pi i}\int_C f(\zeta)\left(\frac{1}{(\zeta - z - h)(\zeta - z)} - \frac{1}{(\zeta - z)^2}\right) d\zeta$$

$$= \frac{1}{2\pi i}\int_C f(\zeta)\frac{h}{(\zeta - z - h)(\zeta - z)^2}\, d\zeta.$$

We must show that $\phi(h)$ has limit 0 as h approaches 0. Since it looks a bit complicated, we as usual estimate it. Since C is closed and bounded, $|f|$ is bounded on C, by M say (using the Boundedness Theorem of *Unit 2*). Let $\delta = r - |z - \alpha|$: then $|\zeta - z| \geqslant \delta$ for $\zeta \in C$, and so $\left|\dfrac{1}{(\zeta - z)^2}\right| \leqslant \dfrac{1}{\delta^2}$ for $\zeta \in C$.

The most troublesome estimate is for $|\zeta - z - h|$, because if $|h|$ is near to δ, $z + h$ may be near to ζ. But let $0 < |h| < \frac{1}{2}\delta$: then $|\zeta - z - h| > \frac{1}{2}\delta$ for $\zeta \in C$. Hence

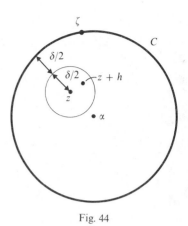

Fig. 44

$$\left| \frac{1}{2\pi i} \int_C f(\zeta) \frac{h}{(\zeta - z - h)(\zeta - z)^2} \, d\zeta \right|$$

$$\leqslant \frac{1}{2\pi} \int_C |f(\zeta)| \frac{|h|}{|\zeta - z - h| \cdot |\zeta - z|^2} \, |d\zeta|$$

$$\leqslant \frac{1}{2\pi} M \frac{|h|}{\frac{1}{2}\delta \cdot \delta^2} \cdot 2\pi r$$

$$= \frac{2Mr}{\delta^3} |h|.$$

Now $\displaystyle \lim_{h \to 0} \frac{2Mr}{\delta^3} |h| = 0$, and so $\displaystyle \lim_{h \to 0} \phi(h) = 0$. Hence f is differentiable at z, and

$$f'(z) = \frac{1}{2\pi i} \int_C \frac{f(\zeta)}{(\zeta - z)^2} \, d\zeta. \quad \blacksquare$$

We now have a formula for $f'(z)$. But in fact this theorem tells us much more, if we look carefully at the proof. We have used the fact that f is analytic only in order to apply Cauchy's Formula at the beginning of the proof: elsewhere we needed just the continuity of f. So we have proved the following result:

If f is continuous on the closed disc consisting of the circle C and its inside D, and for $z \in D, f(z) = \dfrac{1}{2\pi i} \displaystyle\int_C \dfrac{f(\zeta)}{\zeta - z} \, d\zeta$, then f is analytic on D and, in fact, for $z \in D$,

$$f'(z) = \frac{1}{2\pi i} \int_C \frac{f(\zeta)}{(\zeta - z)^2} \, d\zeta$$

(which is what one expects by differentiating under the integral sign).

This result opens up a surprising possibility—since the formula for $f'(z)$ looks very similar to the formula for $f(z)$, perhaps f' is *also* analytic!

This amazing result, unparalleled in real analysis, is what we now prove.

Theorem 6 (Cauchy's Formula for the Second Derivative)

Let f be analytic on some region containing the circle $C = \{z : |z - \alpha| = r\}$ and its inside D. Then f' is analytic on D, and for any z in D,

$$f''(z) = \frac{1}{2\pi i} \int_C \frac{2f(\zeta)}{(\zeta - z)^3} \, d\zeta.$$

(Note that $f'(z) = \displaystyle\int_C H(\zeta, z) \, d\zeta$, where $H(\zeta, z) = \dfrac{1}{2\pi i} \dfrac{f(\zeta)}{(\zeta - z)^2}$, and so $\dfrac{\partial H}{\partial z}(\zeta, z) = \dfrac{1}{2\pi i} \dfrac{2f(\zeta)}{(\zeta - z)^3}$; thus the formula for $f''(z)$ is what we would expect from the remarks above Theorem 5.)

The proof of Theorem 6 has a pattern very similar to the proof of Theorem 5, but the details are slightly more complicated. We suggest that you try to carry it out yourself if you have time before looking at the proof below. But if the details of the proof bewilder you, or you are short of time, leave them until some future reading.

Proof of Theorem 6

Let

$$\phi(h) = \frac{f'(z + h) - f'(z)}{h} - \frac{1}{2\pi i} \int_C \frac{2f(\zeta)}{(\zeta - z)^3} \, d\zeta.$$

We must prove that $\lim_{h \to 0} \phi(h) = 0$. This we do by first simplifying $\phi(h)$ into one integral (using Theorem 5), and then estimating $\phi(h)$ in the manner of Theorem 5.

Now

$$\frac{f'(z+h) - f'(z)}{h} = \frac{1}{2\pi i} \int_C f(\zeta) \frac{1}{h} \left(\frac{1}{(\zeta - z - h)^2} - \frac{1}{(\zeta - z)^2} \right) d\zeta$$

$$= \frac{1}{2\pi i} \int_C f(\zeta) \frac{2(\zeta - z) - h}{(\zeta - z - h)^2 (\zeta - z)^2} d\zeta,$$

since $(\zeta - z)^2 - (\zeta - z - h)^2 = [\zeta - z - (\zeta - z - h)][\zeta - z + (\zeta - z - h)] = h[2(\zeta - z) - h]$. Thus

$$\phi(h) = \frac{1}{2\pi i} \int_C f(\zeta) \left(\frac{2(\zeta - z) - h}{(\zeta - z - h)^2 (\zeta - z)^2} - \frac{2}{(\zeta - z)^3} \right) d\zeta$$

$$= \frac{1}{2\pi i} \int_C f(\zeta) \frac{3h(\zeta - z) - 2h^2}{(\zeta - z - h)^2 (\zeta - z)^3} d\zeta,$$

since $[2(\zeta - z) - h](\zeta - z) - 2[(\zeta - z) - h]^2 = 3h(\zeta - z) - 2h^2$.
Let $\delta = r - |z - \alpha|$, and M be an upper bound for $|f|$ on C. If $0 < |h| < \frac{1}{2}\delta$, then

$$|\zeta - z - h|^2 \geqslant (\tfrac{1}{2}\delta)^2, \quad |\zeta - z|^3 \geqslant \delta^3, \quad \text{and}$$

$$|3h(\zeta - z) - 2h^2| \leqslant 3|h| \cdot 2r + \delta|h|;$$

so that

$$|\phi(h)| \leqslant \frac{1}{2\pi} M \cdot \frac{(6r + \delta)|h|}{(\tfrac{1}{2}\delta)^2 \cdot \delta^3} \cdot 2\pi r = \frac{4M(6r + \delta)r}{\delta^5} |h|,$$

and so $\lim_{h \to 0} \phi(h) = 0$. Hence $f''(z)$ exists and is as required. ∎

The proof above is little more than a routine slog through estimating an integral derived from Cauchy's Formula for the first derivative. But, as we said, the payoff is remarkable!

Theorem 7 (The Analyticity of Derivatives)

Let f be analytic on any region R. Then f' is analytic on R.

Proof

Let $\alpha \in R$. There is certainly $r > 0$ such that the circle $C = \{\zeta : |\zeta - \alpha| = r\}$ and its inside lie in R. By applying Theorem 6 (with $z = \alpha$) to f on the circle C, $f''(\alpha)$ exists. ∎

It is interesting to note that Theorem 7 applies to *any* region R; even though R may not be star, it is "locally star" in the sense that each point of R has a neighbourhood in R which, of course, is a star region, and we apply Cauchy's Formula to these neighbourhoods, not to the whole region.

Theorem 7 is perhaps the first really surprising result of complex analysis, because it directly contradicts one's intuition derived from real analysis. There it is quite easy to construct a differentiable function with non-differentiable derivative. A standard example is the function f given by

$$f(t) = \begin{cases} t^2, & t \geqslant 0 \\ -t^2, & t < 0. \end{cases}$$

Then $f'(t) = 2|t|$, and so f' is not differentiable at 0.

In fact, the property of differentiability is much stronger in complex analysis than in real analysis; we shall see more on this in *Unit 6*. For the moment, though, note that by mathematical induction it follows that if f is analytic on R then all derivatives $f', f'', \ldots, f^{(n)}, \ldots$ of f exist and are analytic on R, that is, f is *infinitely differentiable* on R.

80

So far we have proved that $f^{(n)}$ is analytic on R but we have not found a Cauchy-type formula for $f^{(n)}(z)$. It is, however, obvious that the formula should be:

$$f^{(n)}(z) = \frac{n!}{2\pi i} \int_C \frac{f(\zeta)}{(\zeta - z)^{n+1}} \, d\zeta \, ;$$

for, if $H(\zeta, z) = \dfrac{1}{2\pi i} \dfrac{f(\zeta)}{(\zeta - z)}$, then

$$\frac{\partial^n H}{\partial z^n}(\zeta, z) = \frac{n!}{2\pi i} \frac{f(\zeta)}{(\zeta - z)^{n+1}},$$

and so $f^{(n)}(z)$, that is $\dfrac{d^n}{dz^n}\left(\dfrac{1}{2\pi i} \int_C \dfrac{f(\zeta)}{\zeta - z} \, d\zeta \right)$, should equal

$$\frac{1}{2\pi i} \int_C \frac{\partial^n}{\partial z^n}\left(\frac{f(\zeta)}{\zeta - z} \right) d\zeta, \quad \text{which is} \quad \frac{n!}{2\pi i} \int_C \frac{f(\zeta)}{(\zeta - z)^{n+1}} \, d\zeta.$$

But to *prove* this requires a separate argument. It is presumably clear that to grind out the formula for $f'''(z)$ in the manner of the proof of Theorem 6 would be highly tedious, and the general case for $f^{(n)}(z)$ even worse (though possible). Thus we take a short cut which uses the fact that we *know* from Theorem 7 that $f^{(n)}$ is analytic on R, and so, given any circle C such that C and its inside are contained in R, and any z inside C, then

$$2\pi i f^{(n)}(z) = \int_C \frac{f^{(n)}(\zeta)}{\zeta - z} \, d\zeta.$$

The method we use to transform this integral is *integration by parts*. You may recall Problem 6 in Section 4.2 of *Unit 4* which showed that integration by parts held for complex-valued functions of a real variable. As with many other properties, we can easily prove it next for integrals along smooth arcs and finally for contour integrals (by addition). We end up with the following result, which we state as a lemma, but do not prove.

Lemma (Integration by Parts for Contour Integrals)

Let u and v be functions such that u' and v' are continuous on a contour Γ from a to b. Then

$$\int_\Gamma u'v = uv\Big|_a^b - \int_\Gamma uv'.$$

In particular, if Γ is closed (so that $a = b$),

$$\int_\Gamma u'v = -\int_\Gamma uv'.$$

Now let $u(\zeta) = f^{(n-1)}(\zeta)$, $v(\zeta) = \dfrac{1}{\zeta - z}$, and $\Gamma = C$. Thus

$$\int_C f^{(n)}(\zeta) \frac{1}{\zeta - z} \, d\zeta = \int_C u'v$$

$$= -\int_C uv'$$

$$= -\int_C f^{(n-1)}(\zeta) \frac{-1}{(\zeta - z)^2} \, d\zeta$$

$$= \int_C f^{(n-1)}(\zeta) \frac{1}{(\zeta - z)^2} \, d\zeta.$$

Similarly,

$$\int_C f^{(n-r+1)}(\zeta) \frac{1}{(\zeta - z)^r} \, d\zeta = r \int_C f^{(n-r)}(\zeta) \frac{1}{(\zeta - z)^{r+1}} \, d\zeta \qquad (*)$$

for all r such that $1 \leqslant r \leqslant n$; and so

$$\int_C f^{(n)}(\zeta)\frac{1}{\zeta - z}\,d\zeta = n! \int_C f(\zeta)\frac{1}{(\zeta - z)^{n+1}}\,d\zeta,$$

by applying (∗) for $r = 1, \ldots, n$ in turn. Thus

$$f^{(n)}(z) = \frac{n!}{2\pi i}\int_C \frac{f(\zeta)}{(\zeta - z)^{n+1}}\,d\zeta.$$

What we have done is to prove the following theorem (which includes Theorems 5 and 6).

Theorem 8 (Cauchy's Formula for Derivatives)

Let f be analytic on a region R, C a circle such that C and its inside are contained in R, n any positive integer. Then for any z inside C, f is n-times differentiable at z and

$$f^{(n)}(z) = \frac{n!}{2\pi i}\int_C \frac{f(\zeta)}{(\zeta - z)^{n+1}}\,d\zeta.$$

We have taken a rather roundabout route to this theorem. From Cauchy's Formula for f we derived a formula for f'. We then proved that f' is analytic by deriving Cauchy's Formula for f''. By mathematical induction $f^{(n)}$ was shown to exist and we derived Theorem 8 by integration by parts applied n times.

If we put $n = 0$ in Cauchy's Formula for derivatives, and let $0! = 1$ and $f^{(0)} = f$ as usual, we obtain Cauchy's Formula. We shall normally talk about *Cauchy's Formulas* when n may be zero or positive, *Cauchy's Formula* when n is zero, and *Cauchy's Formula for derivatives* when n is positive.

We can use Cauchy's Formula for derivatives to evaluate contour integrals of the form $\int_C \frac{f(z)}{p(z)}\,dz$ where C is a circle, and p is a polynomial with repeated roots. The techniques are rather similar to those discussed before, and so one example should suffice.

Example 2

Evaluate $\int_C \frac{\sin z}{z^4}\,dz$ where C is the circle $|z| = 2$.

Solution

By Cauchy's Formula for derivatives,

$$\int_C \frac{\sin z}{z^4}\,dz = \frac{2\pi i}{3!}\sin^{(3)}(0) = \tfrac{1}{3}\pi i(-\cos 0) = -\tfrac{1}{3}\pi i.$$

Summary

In this section we have proved Cauchy's Formula, which expresses the values of f *inside* a circle in terms of the values of f *on* the circle—the proof used Cauchy's Theorem and an estimation argument. We then manipulated this formula to give similar formulas for the derivatives of f: first for f', which was fairly easy, and then for f'', which was harder, but gave the important result that f'' *existed* for any analytic f. By a method using integration by parts we were then able to prove versions of Cauchy's Formula for *all* derivatives of f.

The *technique* you have learnt in this section is the evaluation of contour integrals using Cauchy's Formulas.

Self-Assessment Questions

1. Write down Cauchy's Formula (without the various hypotheses).

2. Describe briefly how to prove Cauchy's Formula (Theorem 4). Your answer should include a diagram.

3. Let C be any circle and let z be inside C. Prove that

$$\text{Wnd}(C, z) = \frac{1}{2\pi i} \int_C \frac{d\zeta}{\zeta - z} = 1,$$

where $\text{Wnd}(C, z)$ denotes the winding number of the circle C about z.

4. Evaluate $\int_C \frac{z^2}{z - 1} \, dz$ where C is the circle $|z - 1| = 1$.

5. Write down Cauchy's Formula for the nth derivative.

6. Outline the main steps in the proof of Theorem 8 starting with Cauchy's Formula.

7. Evaluate $\int_C \frac{e^z}{z^2} \, dz$ where C is the circle $|z| = 2$.

8. Given any region R, why is there no function F such that $F'(z) = |z|, z \in R$?

Solutions

1. $f(z) = \frac{1}{2\pi i} \int_C \frac{f(\zeta)}{\zeta - z} \, d\zeta.$

2. (1) We prove that $\int_C \frac{f(\zeta)}{\zeta - z} \, d\zeta = \int_{C'} \frac{f(\zeta)}{\zeta - z} \, d\zeta$, by splitting the region between C and C' into two along the line through α and z, and observing that $\int_{\Gamma_1} \frac{f(\zeta)}{\zeta - z} \, d\zeta = 0$ because Γ_1 lies in a star region, and similarly for Γ_2. (See Fig. 40.)

    (2) Since this holds for all C' centre z we can take the limit of $\int_{C'} \frac{f(\zeta)}{\zeta - z} \, d\zeta$ as the radius ρ of C' approaches 0; this limit can be shown to be $2\pi i f(z)$, by an estimation argument.

3. Apply Theorem 4 (Cauchy's Formula) to the constant function $f(\zeta) = 1$, which is entire.

4. By Cauchy's Formula,

$$\int_C \frac{z^2}{z - 1} \, dz = 2\pi i \cdot 1^2 = 2\pi i.$$

5. $f^{(n)}(z) = \frac{n!}{2\pi i} \int_C \frac{f(\zeta)}{(\zeta - z)^{n+1}} \, d\zeta.$

6. (1) Establish Cauchy's Formula for f'.
    (2) Establish Cauchy's Formula for f'', which shows that f' is analytic.
    (3) By mathematical induction $f^{(n)}$ is analytic.
    (4) Use repeated integration by parts starting from the formula
    $f^{(n)}(z) = \frac{1}{2\pi i} \int_C \frac{f^{(n)}(\zeta)}{\zeta - z} \, d\zeta$ to obtain Cauchy's Formula for $f^{(n)}$.

7. By Cauchy's Formula for derivatives,

$$\int_C \frac{e^z}{z^2} \, dz = \frac{2\pi i}{1!} \exp'(0) = 2\pi i.$$

8. If F did exist it would be analytic on R; and by analyticity of derivatives F' would be analytic, but $F'(z) = |z|$, which is not analytic.

5.6 PROBLEMS

1. This first problem is concerned with routine evaluation of contour integrals using Cauchy's Formulas. Evaluate:

   (i) $\int_C \dfrac{e^{2z}}{(z+1)^4}\, dz$, where C is the circle $|z| = 3$;

   (ii) $\int_C \dfrac{e^{zt}}{z^2 + 1}\, dz$, where C is the circle $|z| = 4$, and $t \in \mathbf{R}$;

   (iii) $\int_C \dfrac{\sin z}{z^2 - z}\, dz$, where C is the circle $|z - 1| = 2$;

   (iv) $\int_C \dfrac{\cos \pi z}{(z^2 - 1)^2}\, dz$, where C is the circle $|z - 1| = 1$.

2. This problem demonstrates that Cauchy's Formulas can be used to evaluate certain real integrals.

   Evaluate $\int_C \dfrac{\exp(z^n)}{z}\, dz$, where C is the circle $|z| = 1$, and n is a positive integer. Hence show that

   $$\int_0^{2\pi} e^{\cos n\theta} \cos(\sin n\theta)\, d\theta = 2\pi,$$

   and

   $$\int_0^{2\pi} e^{\cos n\theta} \sin(\sin n\theta)\, d\theta = 0.$$

3. Here is an estimation involving Cauchy's Formula for derivatives. Show that if f is analytic on a region containing the closed disc $|z| \leqslant r$, then

   $$|f^{(n)}(0)| \leqslant \frac{n!}{2\pi r^n} \int_0^{2\pi} |f(re^{i\theta})|\, d\theta.$$

4. Let f be analytic on a region containing the closed disc $|z - \alpha| \leqslant r$. Show that

   $$f(\alpha) = \frac{1}{2\pi} \int_0^{2\pi} f(\alpha + re^{i\theta})\, d\theta.$$

   (This corollary of Cauchy's Formula is called *Gauss's Mean Value Theorem*. It tells us that the value of f at the centre of a circle is the average, or mean, of the values on the circumference—note that it is quite usual (and even necessary) to express averages as integrals. Gauss's Mean Value Theorem will be used in the proof of the Maximum Principle in *Unit 9*.)

5. Let p_n be defined by

   $$p_n(z) = \frac{1}{n!2^n} \frac{d^n}{dz^n}\big((z^2 - 1)^n\big). \qquad (*)$$

   (a) Show that p_n is a polynomial of degree n.

   (b) Prove that

   $$p_n(z) = \frac{1}{2^{n+1}\pi i} \int_C \frac{(\zeta^2 - 1)^n}{(\zeta - z)^{n+1}}\, d\zeta,$$

   where C is any circle containing z.

   (If you have studied M201, Linear Mathematics, you should recognize $(*)$ as being Rodrigues' Formula for the Legendre polynomials. It is an interesting fact that many of the special functions of analysis can be defined by contour integrals; part (b) gives a definition for the Legendre polynomials.)

6. If you look closely at the proof of Theorem 5, you will see that much less than the analyticity of f and the circularity of C is used. In the light of that try this problem.

Let f be continuous on Γ, where Γ is a contour. Show that the function F given by

$$F(z) = \frac{1}{2\pi i} \int_\Gamma \frac{f(\zeta)}{\zeta - z} \, d\zeta, \qquad z \in \Gamma,$$

is analytic on any region contained in the complement of Γ. (You need not write out a complete proof, but give the main steps.)

Cauchy's Formula for the derivative is a special case of the result of Problem 6. It is also true that Cauchy's Formula *itself* holds for contours other than circles. We shall deal with the most general case in *Unit 9*, but it is very easy to prove many special cases by making only token adjustments to the proof of Theorem 4 and the associated lemma.

7. Let C be a rectangular contour. Let f be analytic on some region containing C and its inside D. Show that for any z in D,

$$f(z) = \frac{1}{2\pi i} \int_C \frac{f(\zeta)}{\zeta - z} \, d\zeta.$$

(Hint: Again use a small circle C' centre z, and note that the proof of Theorem 4 is essentially unchanged, but that the lemma needs some adjustment.)

Clearly a similar result holds for C a triangle, ellipse or any other "geometrically simple" contour enclosing a convex region.

Solutions

1. (i) Since $f(z) = e^{2z}$ is entire and -1 is inside C, we have

$$\int_C \frac{e^{2z}}{(z + 1)^4} \, dz = \frac{2\pi i}{3!} f^{(3)}(-1),$$

$$= \frac{2\pi i}{6} \cdot 8e^{-2}, \quad \text{since } f^{(3)}(z) = 8e^{2z},$$

$$= \frac{8\pi}{3e^2} i.$$

(ii) Let $I = \int_C \frac{e^{zt}}{z^2 + 1} \, dz$; then $I = \int_C \frac{e^{zt}}{2i} \left(\frac{1}{z - i} - \frac{1}{z + i} \right) dz.$

Since $f(z) = e^{zt}$ is entire and i and $-i$ are inside C, we have

$$I = \frac{1}{2i} \cdot 2\pi i (e^{it} - e^{-it})$$

$$= 2\pi i \sin t.$$

(iii) Let $I = \int_C \frac{\sin z}{z^2 - z} \, dz$; then $I = \int_C \sin z \left(\frac{1}{z - 1} - \frac{1}{z} \right) dz.$

Since sin is entire and 1 and 0 are inside C, we have

$$I = 2\pi i (\sin 1 - \sin 0)$$

$$= 2\pi i \sin 1.$$

(iv) Let $I = \int_C \dfrac{\cos \pi z}{(z^2 - 1)^2} \, dz$; then $I = \int_C \dfrac{f(z)}{(z-1)^2} \, dz$, where $f(z) = \dfrac{\cos \pi z}{(z+1)^2}$. Now f

is analytic on $\{z \in \mathbf{C} : z \neq -1\}$· which contains the disc $\{z : |z - 1| \leqslant 1\}$, and so by Cauchy's Formula for f', $I = 2\pi i f'(1)$. Now

$$f'(z) = \frac{(z+1)^2 \cdot (-\pi \sin \pi z) - \cos \pi z \cdot 2(z+1)}{(z+1)^4},$$

and so $f'(1) = \dfrac{0 - (-1)2^2}{2^4} = \dfrac{1}{4}$. Thus $I = \frac{1}{2}\pi i$. (The moral is: you need not always use partial fractions.)

2. We have

$$\int_C \frac{\exp(z^n)}{z} \, dz = 2\pi i e^0 = 2\pi i, \quad \text{by Cauchy's Formula.}$$

Thus

$$2\pi i = \int_0^{2\pi} \frac{\exp\left((e^{i\theta})^n\right)}{e^{i\theta}} \cdot i e^{i\theta} \, d\theta,$$

since C is the path of the arc $\gamma(\theta) = e^{i\theta}$, $\theta \in [0, 2\pi]$. Hence

$$2\pi = \int_0^{2\pi} \exp(e^{in\theta}) \, d\theta$$

$$= \int_0^{2\pi} e^{\cos n\theta} \cdot e^{i \sin n\theta} \, d\theta$$

$$= \int_0^{2\pi} e^{\cos n\theta}[\cos(\sin n\theta) + i \sin(\sin n\theta)] \, d\theta.$$

By equating real and imaginary parts,

$$2\pi = \int_0^{2\pi} e^{\cos n\theta} \cos(\sin n\theta) \, d\theta,$$

and

$$0 = \int_0^{2\pi} e^{\cos n\theta} \sin(\sin n\theta) \, d\theta.$$

3. By Cauchy's Formula for derivatives,

$$|f^{(n)}(0)| = \left| \frac{n!}{2\pi i} \int_C \frac{f(\zeta)}{(\zeta - 0)^{n+1}} \, d\zeta \right|, \quad \text{where } C \text{ is the circle } \{\zeta : |\zeta| = r\}.$$

Since C is the path of the arc $\gamma(\theta) = re^{i\theta}$, $\theta \in [0, 2\pi]$,

$$|f^{(n)}(0)| \leqslant \frac{n!}{2\pi} \int_0^{2\pi} \frac{|f(re^{i\theta})|}{|re^{i\theta}|^{n+1}} |ire^{i\theta}| \, d\theta$$

$$= \frac{n!}{2\pi r^n} \int_0^{2\pi} |f(re^{i\theta})| \, d\theta.$$

4. By Cauchy's Formula,

$$f(\alpha) = \frac{1}{2\pi i} \int_C \frac{f(z)}{z - \alpha} \, dz, \quad \text{where } C \text{ is the circle } |z - \alpha| = r.$$

Since C is the path of the arc $\gamma(\theta) = \alpha + re^{i\theta}$, $\theta \in [0, 2\pi]$, it follows that

$$f(\alpha) = \frac{1}{2\pi i} \int_0^{2\pi} \frac{f(\alpha + re^{i\theta})}{re^{i\theta}} \cdot ire^{i\theta} \, d\theta$$

$$= \frac{1}{2\pi} \int_0^{2\pi} f(\alpha + re^{i\theta}) \, d\theta.$$

5. (a) $(z^2 - 1)^n$ is a polynomial and has degree $2n$, and so $\dfrac{d^n}{dz^n}((z^2 - 1)^n)$ has degree $2n - n = n$. Thus p_n is a polynomial of degree n.

   (b) By Cauchy's Formula for the nth derivative,

$$\frac{n!}{2\pi i} \int_C \frac{(\zeta^2 - 1)^n}{(\zeta - z)^{n+1}} \, d\zeta = \frac{d^n}{dz^n}((z^2 - 1)^n) = n!2^n p_n(z).$$

   The result follows by dividing through by $n!2^n$.

6. It is sufficient to prove that F is differentiable at each point not on Γ. Let $z \notin \Gamma$; take $\delta = d_\Gamma(z)$. See Fig. 45.

Then if $|h| < \delta$,

$$\frac{F(z+h) - F(z)}{h} = \frac{1}{2\pi i} \int_\Gamma f(\zeta) \frac{1}{(\zeta - z - h)(\zeta - z)} \, d\zeta,$$

and so, just as before,

$$\frac{F(z+h) - F(z)}{h} - \frac{1}{2\pi i} \int_\Gamma \frac{f(\zeta)}{(\zeta - z)^2} \, d\zeta = \frac{1}{2\pi i} \int_\Gamma f(\zeta) \frac{h}{(\zeta - z - h)(\zeta - z)^2} \, d\zeta.$$

Let M be an upper bound of $|f|$ on Γ (this exists because f is continuous on Γ). If $0 < |h| < \frac{1}{2}\delta$, then

$$\left| \frac{1}{2\pi i} \int_\Gamma f(\zeta) \frac{h}{(\zeta - z - h)(\zeta - z)^2} \, d\zeta \right|$$
$$\leqslant \frac{1}{2\pi} M \cdot \frac{|h|}{\frac{1}{2}\delta \cdot \delta^2} \ell, \quad \text{where } \ell \text{ is the length of } \Gamma,$$
$$= \frac{M\ell}{\pi\delta^3} |h|.$$

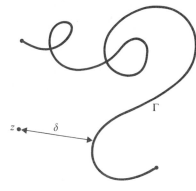

Fig. 45

This has limit zero as h approaches zero, and so

$$F'(z) = \frac{1}{2\pi i} \int_\Gamma \frac{f(\zeta)}{(\zeta - z)^2} \, d\zeta.$$

7. Let C' be a circle inside C with centre z; let g be analytic on some region containing C and its inside, except possibly z. We have to show that $\int_C g = \int_{C'} g$. (If we can do *this*, the proof of Theorem 4 is essentially unchanged.)

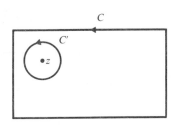

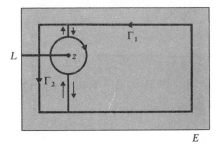

Fig. 46

Join C and C' by lines as shown in Fig. 46, and let Γ_1 and Γ_2 be the two contours shown. Clearly $\left(\int_C - \int_{C'} \right) g = \int_{\Gamma_1} g + \int_{\Gamma_2} g$. We show that $\int_{\Gamma_1} g = \int_{\Gamma_2} g = 0$. If g is not entire, let A be the complement of the region on which g is analytic together with z. Thus d_A is bounded below by some δ on the compact set C. Hence we can find a rectangular region E containing C on which g is analytic (except at z). If g is entire, let $E = C$. Let L be the line as shown. Then $E - L$ is clearly a star region (since E is convex, any point in E is a star, and one on the line of L is clearly a star for $E - L$).

Hence, $\int_{\Gamma_1} g = 0$. Similarly, $\int_{\Gamma_2} g = 0$.

5.7 APPLICATIONS OF CAUCHY'S FORMULAS

We promised you that a host of consequences would flow from Cauchy's Theorem. In fact, most of the flow is channelled through Cauchy's Formulas, which is why we proved these immediately after Cauchy's Theorem. This section deals with such consequences as can be proved without introducing any new concepts. (Consequences involving new concepts will be dealt with in the next few units.)

Our first result is a *converse* of Cauchy's Theorem.

Theorem 9 (Morera's Theorem)*

Let f be a function continuous on a region R, such that for any triangular contour Δ in R with inside contained in R, $\int_\Delta f = 0$. Then f is analytic on R.

Let us reflect on this for a moment, and then the proof should become obvious. We have seen that if f is merely continuous on R, f need not have an anti-derivative. However, the proof of the Antiderivative Theorem is essentially a proof of the following result.

> Let f be a function continuous on a star region D, such that for any triangle Δ in D with inside contained in D, $\int_\Delta f = 0$. Then f has an analytic antiderivative F on D.

Look back at the proof (page 68) if you do not believe this: the above result and the Cauchy–Goursat Theorem add up to the proof of the Antiderivative Theorem. Anyway, if F is analytic on D, so is F' by Theorem 7, the Analyticity of Derivatives—but this is just the function f! To piece together this observation with the above result to give a proof of Morera's Theorem requires one more trick: getting a star region D out of an arbitrary region R. This is easy: since analyticity is a local property, we need only apply the result "locally", that is, in a disc surrounding each point, and a disc is certainly star. Here is the full proof.

Proof of Theorem 9

Let z be a point of R. Then some open disc D centred at z is contained in R. By the above result there is a function F analytic on D such that $F' = f$ on D. But by the Analyticity of Derivatives, F' is analytic on D, that is f is analytic on D. Hence f is analytic on a neighbourhood of z. Since this holds for each $z \in R$, f is analytic on R. ■

Note that Theorem 9 is true for any region—not just star regions.

The main use of Morera's Theorem is as a test for analyticity—a remarkable test, in that it expresses differentiability in terms of continuity and integrability! However, it is extremely useful, since it turns out that differentiability is usually a hard property to verify directly, whereas continuity and integrability are often easier. We shall see an important application of this in *Unit 6*. For the moment, though, here is a fairly straightforward example.

Example 1

Let f be a function continuous on a region R. Suppose that for all $\varepsilon > 0$ there is a function g analytic on R such that
$$|f(z) - g(z)| < \varepsilon, \qquad z \in R.$$

Show that f is analytic on R.

---

* *Giacinto Morera* (1856–1909) discovered this theorem in 1886.

Solution

Let Δ be a triangle in R whose inside is also in R. We shall show that the number $\int_\Delta f$ is zero by the usual method of proving that $\left|\int_\Delta f\right|$ is smaller than every positive number. So let $\varepsilon > 0$. Then there is a function g analytic on R such that $|f(z) - g(z)| < \varepsilon, z \in R$.

Since g is analytic on R, the Cauchy–Goursat Theorem tells us that $\int_\Delta g = 0$,

so that $\int_\Delta f = \int_\Delta (f - g)$. Now we are in business, because

$$\left|\int_\Delta f(z)dz\right| = \left|\int_\Delta [f(z) - g(z)]dz\right|$$

$$\leqslant \int_\Delta |f(z) - g(z)| \cdot |dz|$$

$$\leqslant \ell\varepsilon, \quad \text{where } \ell \text{ is the length of } \Delta.$$

Hence $\left|\int_\Delta f\right| \leqslant \ell\varepsilon$ for all $\varepsilon > 0$, and so $\int_\Delta f = 0$. Since this holds for every such Δ, by Morera's Theorem f is analytic on R.

In loose language, if a continuous function f can be approximated by analytic functions, then f is itself analytic. It may occur to you that the case where g is a Taylor polynomial might be interesting: you would be right, but you will have to wait until *Unit 6*. Also you may have noticed that the approximation was *uniform*, in the sense that ε was independent of z—this kind of approximation is important (see *Units 6* and *11*).

We move on now to a rather different topic, the first of a cluster of results which give severe limitations on the behaviour of analytic functions.

Theorem 10 (Liouville's Theorem)

If f is a bounded entire function, then f is constant.

This may strike you as surprising, since the analogous statement in real analysis is manifestly false. For example, the functions $f(x) = \exp(-x^2)$ and $g(x) = \arctan x$ are bounded and everywhere differentiable, yet very far from constant.

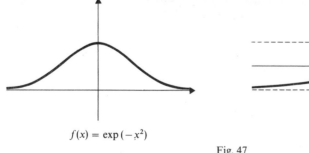

$f(x) = \exp(-x^2)$ $\qquad\qquad$ $g(x) = \arctan x$

Fig. 47

Let us see how one could prove this result, and even motivate it to some extent. An entire function is analytic on the whole complex plane. If f is entire and a is a point in the plane, then $f(a)$ is completely determined by the behaviour of f on any circle C containing a, by Cauchy's Formula

$$f(a) = \frac{1}{2\pi i} \int_C \frac{f(z)}{z - a} dz.$$

If C has a very large radius then it seems that any point b near to a should "look much the same to C as a" in the sense that $f(z)/(z - a)$ and $f(z)/(z - b)$ should be close, for all $z \in C$. But this will hold only if $f(z)$ is not growing as fast as $1/(z - a)$ is shrinking; this certainly holds if f is bounded. Before looking at the proof, spend a few moments seeing if you can now work it out—try to show that $f(a) - f(b) = 0$.

89

Proof of Theorem 10

Let $a, b \in \mathbf{C}$. Then for any circle C containing a and b,

$$f(a) - f(b) = \frac{1}{2\pi i}\int_C \frac{f(z)}{z-a}dz \quad \frac{1}{2\pi i}\int_C \frac{f(z)}{z-b}dz$$

$$= \frac{(a-b)}{2\pi i}\int_C \frac{f(z)}{(z-a)(z-b)}dz.$$

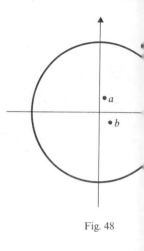

Fig. 48

We now estimate this integral. Let K be such that $|f(z)| \leqslant K$ for all z.

Then if C is the circle $|z| = r$, where $r > \max(|a|, |b|)$,

$$|f(a) - f(b)| \leqslant \frac{|a-b|}{2\pi}\int_C \frac{|f(z)|}{|z-a|\cdot|z-b|}|dz|$$

$$\leqslant \frac{|a-b|}{2\pi}\frac{K}{(r-|a|)(r-|b|)}2\pi r,$$

since $|z-a| \geqslant |z| - |a| = r - |a|$ when $|z| = r$ (and similarly for b),

$$= K|a-b|\frac{r}{(r-|a|)(r-|b|)}.$$

Clearly,

$$\lim_{r\to\infty} K|a-b|\frac{r}{(r-|a|)(r-|b|)} = K|a-b|\lim_{r\to\infty}\frac{r}{r-|a|}\cdot\lim_{r\to\infty}\frac{1}{r-|b|}$$

$$= K|a-b|\cdot 1\cdot 0 = 0.$$

Thus, given any $\varepsilon > 0$ there is r such that

$$|f(a) - f(b)| \leqslant K|a-b|\frac{r}{(r-|a|)(r-|b|)} < \varepsilon;$$

hence $|f(a) - f(b)| < \varepsilon$ for all $\varepsilon > 0$, and so $|f(a) - f(b)| = 0$, that is $f(a) = f(b)$. Since a and b are arbitrary, f is constant. ∎

The next result was mentioned in *Unit 1, Complex Numbers*: its relevance to Liouville's Theorem may come as a shock.

Theorem 11 (The Fundamental Theorem of Algebra)

Let p be a polynomial function of degree at least 1. Then $p(z) = 0$ for at least one z.

We cannot apply Liouville's Theorem directly to p, since p is not bounded, although it is certainly entire. The crux of the proof is the following observation: if $p(z) \neq 0$ for all z, then $1/p$ is entire. Why, though, is $1/p$ then bounded? Roughly, the reasoning is as follows: $|p(z)|$ becomes arbitrarily large for sufficiently large $|z|$, and so $|1/p(z)|$ becomes very close to zero, and hence bounded for sufficiently large $|z|$; for small $|z|$, $|1/p(z)|$ is bounded anyway (by the Boundedness Theorem, because $1/p$ is continuous).

To make such reasoning precise, it is useful (though not essential) to introduce a new type of limit.

Definition

> Let f be a function, and l a complex number. We say that f **has limit l near** ∞, or $f(z)$ **has limit l for large** z, and write $\lim_{z\to\infty} f(z) = l$, if $\lim_{z\to 0} f(1/z) = l$.

From the definition there is an implicit requirement that $f(1/z)$ is defined for all z such that $0 < |z| < \delta$; in other words, $f(z)$ is defined for all z such that $|z| > 1/\delta$, that is the outside of some circle centre the origin.

90

It is easy to check that $\lim\limits_{z \to \infty} f(z) = l$ if and only if for all $\varepsilon > 0$ there is $r > 0$ such that for all z, if $|z| > r$ then $|f(z) - l| < \varepsilon$.

Also $\lim\limits_{z \to \infty} (f + g)(z) = \lim\limits_{z \to \infty} f(z) + \lim\limits_{z \to \infty} g(z)$, and so on.

Using this type of limit we can now prove Theorem 11.

Proof of Theorem 11

Let $p(z) = a_n z^n + a_{n-1} z^{n-1} + \cdots + a_1 z + a_0$, where $n \geq 1$ and $a_n \neq 0$. Since $p(z) = 0 \Leftrightarrow \dfrac{1}{a_n} p(z) = 0$, we can in fact assume that $a_n = 1$. Suppose $p(z) \neq 0$ for all z. Let $f(z) = 1/p(z)$. Then f is entire.

We now show that $\lim\limits_{z \to \infty} f(z) = 0$. Since

$$\frac{p(z)}{z^n} = 1 + \frac{a_{n-1}}{z} + \cdots + \frac{a_1}{z^{n-1}} + \frac{a_0}{z^n},$$

$\lim\limits_{z \to \infty} \dfrac{p(z)}{z^n} = 1$, and so $\lim\limits_{z \to \infty} \dfrac{z^n}{p(z)} = 1$. Thus

$$\lim\limits_{z \to \infty} f(z) = \lim\limits_{z \to \infty} \frac{z^n}{p(z)} \cdot \lim\limits_{z \to \infty} \frac{1}{z^n} = 1 \cdot 0 = 0.$$

Thus there is $r > 0$ such that $|f(z)| < 1$ if $|z| > r$. Since f is continuous on $D = \{z : |z| \leq r\}$, it is bounded on D by some K, by the Boundedness Theorem of *Unit 2*. Hence $|f(z)| \leq K + 1$, for all z. By Liouville's Theorem, f is constant, and so p is constant. This is a contradiction, since p is not constant (for example, note that $p^{(n)}(0) = n! a_n \neq 0$; if p were constant, all its derivatives would be zero.) ∎

Corollary (The Fundamental Theorem of Algebra for Real Analysis)
Let $p(x) = a_n x^n + \cdots + a_1 x + a_0$ be a polynomial with $a_0, \ldots, a_n$ real, $n \geq 1$ and $a_n \neq 0$. Then $p(x) = q(x) r(x)$ for certain polynomials q and r (with real coefficients), where the degree of q is either 1 or 2.

Proof

Consider $p(z)$. Then $p(a) = 0$ for some complex number a. If a is real, let $q(x) = x - a$. Otherwise, $p(\bar{a}) = \overline{p(a)} = 0$: then let

$$q(x) = (x - a)(x - \bar{a}) = x^2 - 2(\text{Re } a)x + |a|^2. \quad ∎$$

By mathematical induction, any polynomial $p(x)$ with real coefficients can be written as a product of linear and quadratic factors.

This corollary is stated in terms of real analysis for functions of one variable, and is an important result—yet no proof in terms of real analysis for functions of one variable is known which is other than hopelessly complicated and unwieldy. Nevertheless, by using complex integration, a rather straightforward proof is obtained. There are other results of real analysis that can be proved in complex analysis in a relatively straightforward way, but which are difficult or impossible to prove in real analysis. See *Unit 8, Singularities*, and *Unit 10, The Calculus of Residues*.

Our final topic fits into this section for two reasons—it is useful in applications of Morera's Theorem, and provides a way of strengthening results such as Cauchy's Formula, that is obtaining the same result with weaker hypotheses. For simplicity we shall discuss it in the context of Cauchy's Theorem. Suppose that f is analytic on a star region R and Γ is a contour not necessarily lying in R but certainly lying in $\bar{R}$, the closure of R (in other words, parts of Γ may lie on the boundary of R). Can we deduce that $\int_\Gamma f = 0$? Note that in order for the integral even to exist we need some assumptions: a reasonable one is that

f is continuous on $\bar{R}$. We can in fact then deduce that $\int_\Gamma f = 0$, and we prove it by approximating Γ by contours which *do* lie in R. The general case is a bit complicated, so here is a special case which nevertheless is fairly typical.

Example 2

Let C be the circle $|z| = R$, and D be the inside of C. Let f be analytic on D and continuous on $\bar{D}$, the closed disc $|z| \leqslant R$. Show that $\int_C f = 0$.

Solution

For $r < R$, the circle $C(r) = \{z : |z| = r\}$ lies in D and so $\int_{C(r)} f = 0$ by Cauchy's Theorem. Thus $\int_C f = \int_C f - \int_{C(r)} f$. As you might now expect, we shall prove that $\int_C f = 0$ by showing that $\left| \int_C f \right|$ is arbitrarily small, for suitable r. Now

$$\left| \int_C f - \int_{C(r)} f \right| = \left| \int_0^{2\pi} f(Re^{i\theta}) \cdot iRe^{i\theta}\, d\theta - \int_0^{2\pi} f(re^{i\theta}) \cdot ire^{i\theta}\, d\theta \right|$$

$$= \left| i \int_0^{2\pi} [f(Re^{i\theta})Re^{i\theta} - f(re^{i\theta})re^{i\theta}]\, d\theta \right|$$

$$\leqslant \int_0^{2\pi} |f(Re^{i\theta})Re^{i\theta} - f(re^{i\theta})re^{i\theta}|\, d\theta.$$

The expression inside this last integral has the form $|f(z)z - f(z')z'|$, that is $|g(z) - g(z')|$ where $g(z) = zf(z)$. The appropriate tool to use in such situations is normally *uniform continuity*. Recall that a function g is uniformly continuous on a set S if for all $\varepsilon > 0$ there is $\delta > 0$ such that for all $z, z' \in S$, if $|z - z'| < \delta$ then $|g(z) - g(z')| < \varepsilon$. You should be able to see how this might help to estimate expressions containing $|g(z) - g(z')|$.

Now g is continuous on the set $\bar{D}$, and so uniformly continuous on $\bar{D}$ (by Problem 4 of Section 2.10 of *Unit 2*).

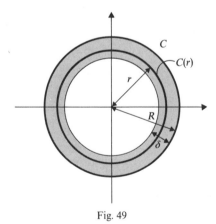

Fig. 49

Let $\varepsilon > 0$, and choose a δ given by uniform continuity of g on $\bar{D}$. Let r be such that $0 < r < R - \delta$: then $|Re^{i\theta} - re^{i\theta}| < \delta$, and so $|f(Re^{i\theta})Re^{i\theta} - f(re^{i\theta})re^{i\theta}| < \varepsilon$ (we are using $g(z) = f(z)z$, with $z = Re^{i\theta}$, $z' = re^{i\theta}$). Hence

$$\left| \int_C f - \int_{C(r)} f \right| \leqslant \int_0^{2\pi} \varepsilon\, d\theta = 2\pi\varepsilon,$$

so that $\left| \int_C f \right| \leqslant 2\pi\varepsilon$. Because this holds for all $\varepsilon > 0$, $\int_C f = 0$.

Notice that we have calculated $\int_\Gamma f$ by approximating Γ, whereas in Example 1 we calculated $\int_\Gamma f$ by approximating f. These techniques will crop up again in later units. But in the next problems section we shall ask you to apply them to obtain a strengthening of Cauchy's Formula.

Note also that C could equally well have been a triangle or rectangle or other similar contour.

Summary

In this section we have proved Morera's Theorem, which provides a test for analyticity involving integration, Liouville's Theorem, on bounded entire functions, and the Fundamental Theorem of Algebra. We ended by discussing how to use uniform continuity to make results such as Cauchy's Theorem apply to contours on the boundary of regions.

Self-Assessment Questions

1. State Morera's Theorem.
2. State Liouville's Theorem.

3. Why is $z \longrightarrow e^{i|z|}$ not entire?

4. What is wrong with the following argument?
   Since $z \longrightarrow \sin z$ is entire and $|\sin z| \leqslant 1$ for all z, $\sin z = \sin 0 = 0$ by Liouville's Theorem.

Solutions

1. See Theorem 9, page 88.

2. See Theorem 10, page 89.

3. Since $|e^{i|z|}| = 1$, and $z \longrightarrow e^{i|z|}$ is not constant, it cannot be entire (otherwise this contradicts Liouville's Theorem).

4. Although $|\sin z| \leqslant 1$ for all *real* z, it is not necessarily true for other z: in fact $|\sin 2i| = \sinh 2 > 1$.

5.8 PROBLEMS

1. Let f be an entire function such that f' is bounded. Show that f is *linear* (that is, there are a and b such that $f(z) = az + b$, $z \in \mathbf{C}$. What can you say when $f^{(n)}$ is bounded?

2. It is possible to strengthen Liouville's Theorem, that is, to prove that f is constant under weaker hypotheses. This problem discusses one way.
   (a) Let f be analytic on a region R, and suppose that $e^f : z \longrightarrow e^{f(z)}$ is constant on R. Show that f is constant on R.
   (b) Let f be an entire function such that $\text{Re}\, f$ is bounded above. Show that f is constant. (Hint: Consider e^f.) What can you say when $\text{Im}\, f$ is bounded above?

3. Let D be the disc $|z - \alpha| < R$. Suppose that f is analytic on D and $|f(z)| \leqslant K$, $z \in D$. Show that for any $n \geqslant 0$,

$$|f^{(n)}(\alpha)| \leqslant \frac{Kn!}{R^n}.$$

(This result is called *Cauchy's Estimate*. It follows from Cauchy's Formulas.)

4. Let f be a non-constant entire function. Show that f cannot have both the periods 1 and i. (Hint: Is f bounded?)

   More generally, it can be shown that if α and β are periods of f then $\text{Arg}\,\alpha = \text{Arg}\,\beta$ or $\text{Arg}\,\alpha = \text{Arg}\,(-\beta)$. Thus all periods of a non-constant periodic entire function f have the same (or opposite) *direction*. Moreover, it can be shown that every period α of f has the form $n\beta$ for some nonzero integer n, where β is a fixed period of f, called the *fundamental period* of f. We shall not *prove* this fact because it has nothing to do with complex analysis, being merely the translation of the corresponding result in real analysis; however, it is worth remembering.

5. (i) Suppose that f is an entire function and k is a positive real number such that $|f(z)| \geqslant k$ for all z. Show that f is constant.
   (ii) Suppose that f is an entire function, α is a complex number and k is a positive real number such that $|f(z) - \alpha| \geqslant k$ for all z. Show that f is constant.

6. Let f be analytic on the inside D of the circle $C = \{z : |z| = R\}$ and continuous on $\bar{D}$, the closure of D, and let $\alpha \in D$. Show that

$$f(\alpha) = \frac{1}{2\pi i} \int_C \frac{f(z)}{z - \alpha}\, dz.$$

(Hint: See Example 2 of Section 5.7.)

Solutions

1. Clearly f' is entire (by Analyticity of Derivatives). Since f' is also bounded there is $a \in \mathbf{C}$ such that $f'(z) = a$, $z \in \mathbf{C}$, by Liouville's Theorem. By integrating, $f(z) = az + b$, for some constant b.

   If $f^{(n)}$ is bounded, then $f^{(n)}(z) = a_n$ for some constant a_n, since $f^{(n)}$ is also entire. Hence, by integrating n times,

$$f(z) = a_n z^n + a_{n-1} z^{n-1} + \cdots + a_0,$$

for certain constants $a_{n-1}, \ldots, a_0$. Thus f is a polynomial of degree n.

2. (a) Suppose that $e^{f(z)} = k$, $z \in R$. Let z_0 be some point of R, and let $\alpha = f(z_0)$. Then $e^\alpha = k$, so that α is a logarithm of k. We shall show that $f(z) = \alpha$, $z \in R$. The proof uses the key fact that the logarithms of k are isolated, in the sense that if α and β are logarithms of k and $\alpha \neq \beta$ then $|\alpha - \beta| \geqslant 2\pi$ (since $\beta = \alpha + 2n\pi i$ for some nonzero integer n). We use a connectedness argument.

Consider $G_1 = \{z \in R : f(z) = \alpha\}$, and $G_2 = \{z \in R : f(z) \neq \alpha\}$. Then $G_1 \cup G_2 = R$ and $G_1 \cap G_2 = \varnothing$. Clearly G_2 is open. But G_1 is also open, since $G_1 = f^{-1}(D) \cap R$, where $D = \{w : |w - \alpha| < 2\pi\}$, and f is continuous on R. Since R is connected and $G_1 \neq \varnothing$, $G_2 = \varnothing$, so that $R = G_1$; thus $f(z) = \alpha$ for *all* $z \in R$.

(b) Let $g(z) = e^{f(z)}$. Then g is entire. Suppose that $\operatorname{Re} f(z) \leqslant K$ for all z. Then

$$|g(z)| = |e^{\operatorname{Re} f(z)}| \cdot |e^{i \operatorname{Im} f(z)}| = e^{\operatorname{Re} f(z)} \cdot 1 \leqslant e^K$$

because $\exp : \mathbf{R} \longrightarrow \mathbf{R}$ is increasing. Hence g is entire and bounded. Thus by Liouville's Theorem, g is constant, with value α, say. Hence $e^{f(z)} = \alpha$ for all z. Thus, by part (a), f is constant.

Note that $\operatorname{Im} f = \operatorname{Re}(-if)$. Thus if $\operatorname{Im} f$ is bounded above, then $-if$ is constant, and so f is constant.

3. Let $r < R$, and $C(r)$ be the circle $\{z : |z - \alpha| = r\}$. Then by Cauchy's Formulas applied to $C(r)$, we have

$$f^{(n)}(\alpha) = \frac{n!}{2\pi i} \int_{C(r)} \frac{f(z)}{(z - \alpha)^{n+1}} \, dz.$$

Hence

$$|f^{(n)}(\alpha)| \leqslant \frac{n!}{2\pi r^{n+1}} \int_{C(r)} |f(z)| \cdot |dz|$$

$$\leqslant \frac{n!}{2\pi r^{n+1}} K \cdot 2\pi r$$

$$= \frac{Kn!}{r^n}, \quad \text{since } |z - \alpha| = r \text{ for } z \in C(r).$$

Since this holds for *all* $r < R$, we can take limits: thus

$$|f^{(n)}(\alpha)| \leqslant \lim_{r \to R^-} \frac{Kn!}{r^n} = \frac{Kn!}{R^n}.$$

(Note that we cannot yet use Cauchy's Formula for $C(r)$, as $C(r)$ does not lie in D. However, see Problem 6.)

4. Let R be the closed square with vertices $0, 1, i, 1 + i$. Since R is closed and bounded, f is bounded, by some K, say, on R. Now assume that 1 and i are periods of f; thus $f(z + m + ni) = f(z)$ for any $m, n \in \mathbf{Z}$. Given $z' \in \mathbf{C}$,

$$|f(z')| = |f(z + m + ni)| \text{ for certain } m, n \in \mathbf{Z}, \text{ and } z \in R,$$

$$= |f(z)| \leqslant K.$$

Hence f is bounded by K on $\mathbf{C}$, and so f is constant (by Liouville's Theorem). This is a contradiction and so f does not have periods 1 and i.

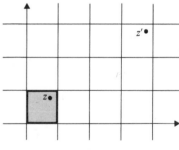

Fig. 50

5. (i) Let $g(z) = 1/f(z)$. Since $f(z) \neq 0$, g is entire. Also $|g(z)| \leqslant 1/k$. Thus g is constant by Liouville's Theorem. Hence f is constant.

(ii) From part (i) the function $z \longrightarrow f(z) - \alpha$ is constant; therefore f is constant.

6. It is convenient to introduce the function $F(z) = \dfrac{1}{2\pi i} \dfrac{f(z)}{z - \alpha}$. What we have to prove is that $f(\alpha) = \displaystyle\int_C F(z) \, dz$. Now for $r < R$, $f(\alpha) = \displaystyle\int_{C(r)} F(z) \, dz$ by Cauchy's Formula, where $C(r)$ is the circle $\{z : |z| = r\}$. We shall show that

$$\int_C F(z) \, dz = \lim_{r \to R^-} \int_{C(r)} F(z) \, dz.$$

This follows much as in Example 2. Let $|\alpha| < R_0 < R$. If $R_0 < r < R$, then

$$\left| \int_C F - \int_{C(r)} F \right| \leqslant \int_0^{2\pi} |F(Re^{i\theta})Re^{i\theta} - F(re^{i\theta})re^{i\theta}| \, d\theta.$$

Now $z \longrightarrow F(z)z$ is continuous on the annulus $A = \{z : R_0 \leqslant |z| \leqslant R\}$ and so uniformly continuous on A. Let $\varepsilon > 0$, and δ be given by uniform continuity. Then, if $R > r > R - \delta$, we have

$$\left| \int_C F - \int_{C(r)} f \right| \leqslant 2\pi\varepsilon.$$

The result follows.

95

5.9 TECHNIQUES OF INTEGRATION

Let f be analytic on a star region R and Γ be a contour in R. Then by the Antiderivative Theorem, f has at least one antiderivative F on R. This opens up the possibility of evaluating $\int_\Gamma f(z)\,dz$ by calculating $F(\beta) - F(\alpha)$, where α and β are the endpoints of Γ. However, as in real analysis, we may not be able to find any antiderivative of f given by a pleasant formula—note that those given by the Antiderivative Theorem are not really any help in actually *calculating* $\int_\Gamma f(z)\,dz$. But in many cases, if f is given by a pleasant formula, one's knowledge of differentiation may lead one to an antiderivative of f. What we shall do now is discuss some systematic techniques to help such a search—these are the analogues of integration by parts and integration by substitution, familiar to you from real analysis.

Given that f is analytic on a region R, we shall use the traditional notation $\int f$ or $\int f(z)\,dz$, to mean "an antiderivative of f on R"—usually R is understood from the context. Thus $\int f = F$ just means $F' = f$ on R—we sometimes call F an *integral* of f. Sometimes we are even less precise and write $\int f(z)\,dz = F(z)$ for $\int f = F$. This lack of precision is justified when we are *searching* for antiderivatives, when technique and convenient notation are the main considerations. When it comes to *proofs* we are more careful.

If f has an antiderivative on R, then contour integrals of f in R are *path-independent*, in the sense that given any two points α, β in R, then the value of $\int_\Gamma f(z)\,dz$ is independent of the choice of contour Γ in R from α to β—in this case we write $\int_\alpha^\beta f(z)\,dz$ for this constant value. Since $\int_\alpha^\beta f(z)\,dz = F(\beta) - F(\alpha)$ where $\int f = F$, calculation of contour integrals is very dependent on finding integrals.

Here are some *standard integrals*. You can check them by differentiation.

$$\int 1\,dz = z.$$

$$\int z^n\,dz = \frac{z^{n+1}}{n+1}, \quad n \text{ a positive integer.}$$

$$\int e^z\,dz = e^z.$$

$$\int \sin z\,dz = -\cos z.$$

$$\int \cos z\,dz = \sin z.$$

$$\int \sinh z\,dz = \cosh z.$$

$$\int \cosh z\,dz = \sinh z.$$

Note that in all these cases the appropriate region R is $\mathbf{C}$. The most interesting integral not of this kind is

$$\int \frac{1}{z}\,dz = \operatorname{Log} z, \quad \text{where the region } R \text{ is } \mathbf{C} - \{t \in \mathbf{R} : t \leqslant 0\}.$$

(In *Unit 3*, we proved that $\text{Log}'(z) = 1/z$, $z \in R$. If R is not $\mathbf{C} - \{t \in R : t \leqslant 0\}$, then $\int \frac{1}{z} dz$ will be a different function.)

Suppose that $F' = f$ and $G' = g$ on R. Then $(F + G)' = f + g$ on R. In the terminology introduced above this is written $F + G = \int (f + g)$, that is, $\int (f + g) = \int f + \int g$. You can read this "equation" as the following "recipe": to find an antiderivative of $f + g$ add an antiderivative of f to an antiderivative of g. Similarly $\int \lambda f = \lambda \int f$, for $\lambda \in \mathbf{C}$. But here is our first *important* technique, integration by parts.

Theorem 12 (Integration by Parts)

Let f and g be analytic on a region R. Then

(i) $\displaystyle \int f \cdot g' = f \cdot g - \int f' \cdot g$ on R;

(ii) $\displaystyle \int_{\alpha}^{\beta} f(z)g'(z)\, dz = f(z)g(z)\big|_{\alpha}^{\beta} - \int_{\alpha}^{\beta} f'(z)g(z)\, dz$, for $\alpha, \beta \in R$.

Note that, unlike real analysis, we do not need to add the hypotheses that f' and g' are continuous, since by Analyticity of Derivatives, f' and g' are in fact analytic. Note also that if R is not a star region then the antiderivatives may not exist, so that one must interpret the "equations" (1) and (2) carefully, rather in the way that one interprets equations involving limits—that is, if the right-hand side exists, so does the left.

Proof of Theorem 12

(i) Let $\int f' \cdot g = H$. Then $H' = f' \cdot g$. But, since $(f \cdot g)' = f' \cdot g + f \cdot g'$,

$$f \cdot g' = (f \cdot g)' - f' \cdot g = (f \cdot g - H)',$$

that is

$$\int f \cdot g' = f \cdot g - H = f \cdot g - \int f' \cdot g.$$

(ii) By the Fundamental Theorem,

$$\int_{\alpha}^{\beta} f(z)g'(z)\, dz = f(z)g(z)\big|_{\alpha}^{\beta} - \int_{\alpha}^{\beta} f'(z)g(z)\, dz. \quad \blacksquare$$

Of course, if R *is* a star region then $f' \cdot g$ and $f \cdot g'$ have antiderivatives; and this is the most usual case.

Example 1

Evaluate on suitable regions

(i) $\displaystyle \int z e^z \, dz$ and   (ii) $\displaystyle \int_{1}^{i} \text{Log}\, z \, dz$.

Solution

(i) $\displaystyle \int z e^z \, dz = z e^z - \int 1 \cdot e^z \, dz = z e^z - e^z$.

We let R be $\mathbf{C}$, which is a star region: then we integrate $z \longrightarrow e^z$ and differentiate $z \longrightarrow z$.

(ii) $\displaystyle\int_1^i \mathrm{Log}\, z\, dz = \int_1^i 1 \cdot \mathrm{Log}\, z\, dz = z\,\mathrm{Log}\, z\big|_1^i - \int_1^i z \cdot \frac{1}{z}\, dz$

$$= (z\,\mathrm{Log}\, z - z)\big|_1^i$$

$$= 1 - \frac{\pi}{2} - i.$$

We let R be $\mathbf{C} - \{t \in \mathbf{R} : t \leqslant 0\}$, which again is a star region; then we integrate $z \longrightarrow 1$ (a common trick) and differentiate Log.

Theorem 13 (Integration by Substitution)

Let f be analytic on the region S and let $g : R \longrightarrow S$ be analytic on the region R. Then

(i) $\displaystyle\int (f \circ g) \cdot g' = \left(\int f\right) \circ g$ on R,

(ii) $\displaystyle\int_\alpha^\beta f(g(z))g'(z)\, dz = \int_{g(\alpha)}^{g(\beta)} f(w)\, dw$, for $\alpha, \beta \in R$.

Proof

(i) Note first that g' is also analytic on R. Now let $\int f = F$ (integrating on S).

Then $(f \circ g) \cdot g' = (F' \circ g) \cdot g' = (F \circ g)'$, and so (integrating on R)

$$\int (f \circ g) \cdot g' = F \circ g = \left(\int f\right) \circ g.$$

(ii) Hence, by the Fundamental Theorem,

$$\int_\alpha^\beta f(g(z))g'(z)\, dz = F \circ g\big|_\alpha^\beta = F\big|_{g(\alpha)}^{g(\beta)} = \int_{g(\alpha)}^{g(\beta)} f(w)\, dw.$$

(Note that for convenience we have used the form $f\big|_a^b$ rather than $f(z)\big|_a^b$.) ∎

Example 2

Evaluate on suitable regions

(i) $\displaystyle\int z \sin z^2\, dz$ and (ii) $\displaystyle\int_0^\pi \frac{\sin z}{\cos^2 z}\, dz.$

Solution

(i) $\displaystyle\int z \sin z^2\, dz = \int \tfrac{1}{2} \sin w\, dw$, where $w = z^2$, $dw = 2z\, dz$,

$$= -\tfrac{1}{2} \cos w = -\tfrac{1}{2} \cos z^2.$$

This is the normal way of setting out an integration by substitution. In the notation of Theorem 13 we have $f(w) = \sin w$, $g(z) = z^2$, $R = S = \mathbf{C}$.

(ii) $\displaystyle\int_0^\pi \frac{\sin z}{\cos^2 z}\, dz = \int_1^{-1} -\frac{1}{w^2}\, dw,$

where $w = \cos z$, $dw = -\sin z\, dz$, $\cos 0 = 1$, $\cos \pi = -1$,

$$= \frac{1}{w}\Big|_1^{-1} = -1 - 1 = -2.$$

In this case $f(w) = -1/w^2$, $S = \{z : z \neq 0\}$, and $g(z) = \cos z$, $z \in R$, where $R = \{z : \cos z \neq 0\} = \mathbf{C} - \{(2n+1)\frac{\pi}{2} : n \in \mathbf{Z}\}$. Neither R nor S is a star region, but nevertheless f has an antiderivative on R, and so Theorem 13 tells us that $(f \circ g) \cdot g'$ does. Note also that $\int_0^{\pi} \dfrac{\sin x}{\cos^2 x}\, dx$ does not exist in real analysis since $\cos \dfrac{\pi}{2} = 0$; by going into the complex plane we can avoid troublesome points such as $\pi/2$. (We deal with such integrals in *Unit 10*.)

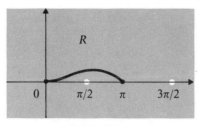

Fig. 51

You should already know how to integrate in real analysis. The techniques in complex analysis are very similar, except that it is harder to carry out rigorous arguments, because the domains of functions are more complicated and fewer functions are one–one. The best thing to do in such cases is to hack away, applying the techniques in the manner you know from real analysis, and check the answer you get by differentiation on a suitable region. The aim, after all, is to find a correct answer efficiently—the proof of the integration is in the differentiation!

Summary

In this section we have discussed techniques of integration, and proved theorems on integration by parts and integration by substitution.

Self-Assessment Question

Find

(i) $\displaystyle\int (\sin z + 2 \cos z)\, dz,$

(ii) $\displaystyle\int z \cosh z\, dz,$

(iii) $\displaystyle\int \exp(e^z) e^z\, dz.$

Solution

(i) $\displaystyle\int (\sin z + 2 \cos z)\, dz = -\cos z + 2 \sin z.$

(ii) $\displaystyle\int z \cosh z\, dz = z \sinh z - \int 1 \cdot \sinh z\, dz$

$\qquad\qquad = z \sinh z - \cosh z.$

(iii) $\displaystyle\int \exp(e^z) e^z\, dz = \exp(e^z)$ (let $w = e^z$).

These integrals hold on any region.

5.10 PROBLEMS

1. Calculate $\int_0^{\pi i} z^2 e^z dz$.

2. Find the following integrals.

   (i) $\int \tan z\, dz$.

   (ii) $\int \dfrac{dz}{z^2 + 1}$. (Hint: Partial fractions.)

   (iii) $\int \dfrac{dz}{1 + e^z}$.

3. (i) Evaluate $\int_0^{2i} \dfrac{dz}{z^2 + 1}$.

   (ii) How would you evaluate $\int_{-1-i}^{-1+i} \dfrac{dz}{z^2 + 1}$?

Solutions

1. $\displaystyle\int_0^{\pi i} z^2 e^z\, dz = z^2 e^z\Big|_0^{\pi i} - \int_0^{\pi i} 2z e^z\, dz$

   $\qquad\qquad = \pi^2 - \left(2z e^z\Big|_0^{\pi i} - \int_0^{\pi i} 2e^z\, dz \right)$

   $\qquad\qquad = \pi^2 + 2\pi i + 2e^z\Big|_0^{\pi i}$

   $\qquad\qquad = (\pi^2 - 4) + 2\pi i.$

   Thus given any contour Γ joining 0 to πi, $\displaystyle\int_\Gamma z^2 e^z\, dz = (\pi^2 - 4) + 2\pi i.$

2. (i) $\displaystyle\int \tan z\, dz = -\int \dfrac{-\sin z}{\cos z}\, dz = -\text{Log}(\cos z).$

   Check: $\dfrac{d}{dz}(-\text{Log}(\cos z)) = -\dfrac{-\sin z}{\cos z} = \tan z$, except when $\cos z$ is real and non-positive. Fig. 52 illustrates the set of z for which the result holds.

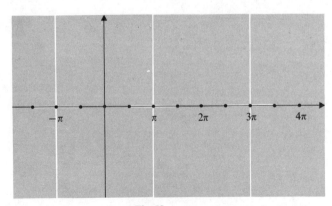

Fig. 52

   (ii) $\displaystyle\int \dfrac{dz}{z^2 + 1} = \int \dfrac{1}{2i}\left(\dfrac{1}{z - i} - \dfrac{1}{z + i} \right) dz$

   $\qquad\qquad = \dfrac{1}{2i}[\text{Log}(z - i) - \text{Log}(z + i)].$

Check: $\frac{d}{dz}\left(\frac{1}{2i}[\text{Log}\,(z-i) - \text{Log}\,(z+i)]\right) = \frac{1}{z^2+1}$ if $z = x + iy$ where $x > 0$ or

$y \neq \pm i$. Fig. 53 illustrates the set of such z.

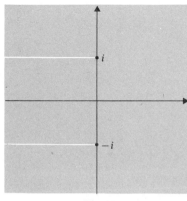

Fig. 53

(iii) $\displaystyle\int \frac{dz}{1+e^z} = \int \frac{1}{1+w}\cdot\frac{1}{w}\,dw$, where $z = \text{Log}\,w$, $dz = \frac{1}{w}\,dw$,

$$= \int\left(\frac{1}{w} - \frac{1}{w+1}\right)dw$$

$$= \text{Log}\,w - \text{Log}\,(w+1)$$

$$= z - \text{Log}\,(e^z + 1).$$

Check: $\frac{d}{dz}(z - \text{Log}\,(e^z + 1)) = 1 - \frac{e^z}{e^z+1} = \frac{1}{1+e^z}$, except when e^z is real and

not greater than -1. Fig. 54 illustrates the set of z for which the result holds.

3. (i) $\displaystyle\int_0^{2i} \frac{dz}{z^2+1} = \frac{1}{2i}[\text{Log}\,(z-i) - \text{Log}\,(z+i)]\Big|_0^{2i}$, by Problem 2 (ii),

$$= \frac{1}{2i}[\text{Log}\,i - \text{Log}\,3i - \text{Log}\,(-i) + \text{Log}\,i]$$

$$= \frac{1}{2i}\left[2i\frac{\pi}{2} - \left(\log 3 + \frac{i\pi}{2}\right) - \left(-i\frac{\pi}{2}\right)\right],$$

$$\text{since } \text{Log}\,ki = \begin{cases} \log k + i\dfrac{\pi}{2}, & k > 0 \\[2mm] \log(-k) - i\dfrac{\pi}{2}, & k < 0, \end{cases}$$

$$= \frac{\pi}{2} + \frac{1}{2}i\log 3.$$

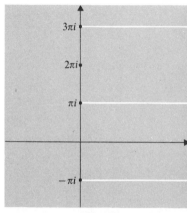

Fig. 54

Given any contour Γ in the region of Fig. 53 from 0 to $2i$, we have

$$\int_\Gamma \frac{dz}{z^2+1} = \frac{\pi}{2} + \frac{1}{2}i\log 3.$$

(ii) However, to evaluate $\displaystyle\int_{-1-i}^{-1+i} \frac{dz}{z^2+1}$, we must choose a different branch of the

logarithm. Let $L(z) = \log|z| + iA(z)$, where $A(z)$ is the argument of z lying in $(0, 2\pi)$: then the domain of L is $\mathbf{C} - \{t \in \mathbf{R}: t \geqslant 0\}$. The domain of $z \longrightarrow L(z-i) - L(z+i)$ is shown in Fig. 55. (Note that this branch of the logarithm was discussed in Section 3.6 of *Unit 3*.)

Then

$$\int_{-1-i}^{-1+i} \frac{dz}{z^2+1} = \frac{1}{2i}[L(z-i) - L(z+i)]\Big|_{-1-i}^{-1+i}.$$

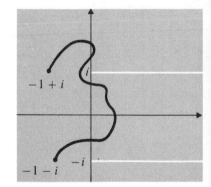

Fig. 55

This branch of the logarithm will enable us to integrate along any contour from $-1-i$ to $-1+i$ in the region shown in Fig. 55.

101

Unit 6 Taylor Series

Conventions

Before working through this text make sure you have read *A Guide to the Course: Complex Analysis.*

References to units of other Open University courses in mathematics take the form:

Unit M100 13, Integration II.

The set book for the course M231, Analysis, is M. Spivak, *Calculus*, paperback edition (W. A. Benjamin/Addison-Wesley, 1973). This is referred to as:

Spivak.

Optional Material

This course has been designed so that it is possible to make minor changes to the content in the light of experience. You should therefore consult the supplementary material to discover which sections of this text are not part of the course in the current academic year.

6.0 INTRODUCTION

In this unit we take an apparent change of direction. After spending most of our time so far discussing differentiation and integration, we now turn our attention to power series. But such a move is not as contrary as it may seem. The effectiveness of the work we have done so far is limited unless we have a good supply of functions to hand. Power series give us a good way of specifying functions. Furthermore, we shall see in this unit that every function analytic on a disc can be represented by power series and this representation is a useful analytic device. Thus this unit is a natural successor to the earlier units in two ways—it extends our ability to specify functions and it extends our analytical tool kit. Furthermore, the major analytical techniques required to develop the topics of this unit are those obtained in *Unit 5, Cauchy's Theorem I*, and so this unit can be regarded as a major application of the results of *Unit 5*.

You probably recall that sufficiently well-behaved real functions can be represented by power series, their Taylor series. The same is true of complex functions and in this context "sufficiently well-behaved" means analytic.

Section 6.1 presents a review of some results about real series, and a brief account of the theory of complex series. This section is rather single-minded in the sense that it is included to serve the main purpose of the unit—a discussion of complex *power* series—rather than an introduction to the theory of *general* complex series. We shall discuss more general complex series in *Unit 11, Analytic Functions*.

The set of points at which a power series converges forms the domain of the function defined by the series. In Section 6.3 we see first of all that this set has a special form—a disc. This result is analogous to the real analysis result in which power series are shown to converge on an interval. However, we can say more about a function defined by a complex power series than one defined by a real power series. In particular, we shall see that the function is analytic and it can be differentiated just by differentiating the series term by term.

In Section 6.5 we turn the problem round—given a function analytic on a disc we find out how to construct a power series to represent it. This is the Taylor series. This representation takes one of two closely related forms—an infinite series, or a polynomial approximation together with a remainder term. The fact that every function analytic on a disc has a Taylor representation means that Taylor series can be usefully employed to investigate the properties of such functions. As an example of this we show in Section 6.7 that even less information is required to determine an analytic function uniquely than one would expect from Cauchy's Formula, which tells us that an analytic function is determined uniquely by its values on a circle. In fact, we shall show that a knowledge of the values of a function on a convegent sequence of points with limit in the domain of the function is sufficient to determine the function uniquely in a region.

Television and Radio

In the fourth television programme associated with the course we shall illustrate the equivalence between analytic functions and Taylor series, and in the third radio programme we shall consider examples of Taylor series.

6.1 CONVERGENCE OF COMPLEX SERIES

Real Series: Reminders

We have collected together several definitions and results to do with series of real numbers, which will be needed in this unit. Read them through for revision, and while you are doing so, try to decide whether similar definitions and results will work for series of complex numbers, and what alterations, if any, will have to be made. (A full discussion of real series is contained in *Unit M231 14, Sequences and Series*, which is based on Chapters 21 and 22 of **Spivak**.)

An **infinite sequence** of real numbers is a function from the natural numbers **N** to the real numbers **R**. We denote the sequence $n \longrightarrow a_n$ by $\{a_n\}$.

Definition

> A sequence of real numbers $\{a_n\}$ **converges to** l if for every $\varepsilon > 0$, there is a natural number N such that, for all natural numbers n,
>
> $$\text{if } n > N, \text{ then } |a_n - l| < \varepsilon.$$
>
> We then write $\lim_{n \to \infty} a_n = l$.

A sequence $\{a_n\}$ is a **Cauchy sequence** if for every $\varepsilon > 0$ there is a natural number N such that, for all m and n,

$$\text{if } m, n > N, \text{ then } |a_n - a_m| < \varepsilon.$$

A sequence converges if and only if it is a Cauchy sequence.

Definition

> The sequence of real numbers $\{a_n\}$ is **summable** if the sequence $\{s_n\}$ converges, where
>
> $$s_n = \sum_{k=1}^{n} a_k.$$
>
> The (infinite) series of real numbers $\sum_{n=1}^{\infty} a_n$ **converges to the sum** s if $\lim_{n \to \infty} s_n$ exists and equals s.

The symbols $\sum_{n=1}^{\infty} a_n$ are used to denote the series itself and the sum if it exists. The intended sense is usually clear from the context.

Some Frequently Encountered Series

The geometric series, $\sum_{n=0}^{\infty} r^n$, converges if $|r| < 1$, diverges (that is, does not converge) if $|r| \geqslant 1$.

The series, $\sum_{n=1}^{\infty} \frac{1}{n^p}$, where p is a real number, converges if $p > 1$, diverges if $p \leqslant 1$.

Tests for Convergence of Series of Non-negative Terms

The comparison test: If $0 \leqslant a_n \leqslant b_n$ for all n, and the series $\sum_{n=1}^{\infty} b_n$ converges, then so does the series $\sum_{n=1}^{\infty} a_n$.

The ratio test: If $a_n > 0$ for all n, and $\lim_{n \to \infty} a_{n+1}/a_n$ exists and equals r, say, then the series $\sum_{n=1}^{\infty} a_n$ converges if $r < 1$ and diverges if $r > 1$. No conclusion may be drawn if $r = 1$, or if the limit does not exist.

The Cauchy Criterion for Convergence of Series

The series $\sum_{n=1}^{\infty} a_n$ converges if and only if, given $\varepsilon > 0$, there is some natural number N such that

$$|a_{n+1} + a_{n+2} + \cdots + a_{n+k}| < \varepsilon \quad \text{whenever } n \geqslant N \text{ and } k \geqslant 1.$$

In particular, if $\sum_{n=1}^{\infty} a_n$ converges, then $\lim_{n \to \infty} a_n = 0$. (But $\lim_{n \to \infty} a_n = 0$ is not a sufficient condition for convergence.)

Definition

> The series of real numbers $\sum_{n=1}^{\infty} a_n$ is **absolutely convergent** if the series $\sum_{n=1}^{\infty} |a_n|$ is convergent.

An absolutely convergent series is convergent, but a convergent series need not necessarily be absolutely convergent.

The Cauchy Product

If $\sum_{n=1}^{\infty} a_n$ and $\sum_{n=1}^{\infty} b_n$ converge absolutely and $c_n = a_1 b_n + a_2 b_{n-1} + \cdots + a_n b_1$, then $\sum_{n=1}^{\infty} c_n$ converges, and $\sum_{n=1}^{\infty} c_n = \left(\sum_{n=1}^{\infty} a_n \right) \cdot \left(\sum_{n=1}^{\infty} b_n \right)$. The series $\sum_{n=1}^{\infty} c_n$ is called the *Cauchy product* of the series $\sum_{n=1}^{\infty} a_n$ and $\sum_{n=1}^{\infty} b_n$.

Convergence of Complex Series

The basic definitions concerning complex series and sequences are almost identical in appearance with those for real series. Our main interest in this unit is *power series*, in which the variable can take complex values. In this section, therefore, we shall go through the essential definitions for complex series but we shall not try to be comprehensive in our treatment: our aim is to equip ourselves with just enough information to tackle power series in Section 6.3.

An **infinite sequence** of complex numbers is a function from the natural numbers **N** to the complex numbers **C**. As with real sequences, we use the notation $\{a_n\}$ for the sequence $n \longrightarrow a_n$. The definition of convergence is formally just the same as for real sequences.

Definition

> A sequence of complex numbers $\{a_n\}$ **converges to** l if for every real $\varepsilon > 0$, there is a natural number N such that, for all natural numbers n,
>
> if $n > N$, then $|a_n - l| < \varepsilon$.
>
> We then write $\lim_{n \to \infty} a_n = l$.

107

Although the definition looks the same as that for real sequences, the interpretation is different in the sense that all terms 'eventually' lie in the open disc $\{z: |z - l| < \varepsilon\}$, instead of the interval $(l - \varepsilon, l + \varepsilon)$. (In the definition we have stated *explicitly* that ε is real; hereinafter ε is always real.)

The convergence of complex sequences can be considered directly in terms of the convergence of real sequences, as a result of the following theorem.

Theorem 1

Let $a_n = \alpha_n + i\beta_n$ and $l = \alpha + i\beta$, where $\alpha_n, \beta_n, \alpha$ and β are real numbers. Then

$$\lim_{n \to \infty} a_n = l \text{ if and only if } \lim_{n \to \infty} \alpha_n = \alpha \text{ and } \lim_{n \to \infty} \beta_n = \beta.$$

You are asked to prove this theorem in Problem 1 of the next section.

Just as real series are discussed in terms of real sequences, so complex series are discussed in terms of complex sequences.

Definition

The sequence of complex numbers $\{a_n\}$ is **summable** if the sequence $\{s_n\}$ converges, where

$$s_n = \sum_{k=1}^{n} a_k.$$

The (infinite) series of complex numbers $\sum_{n=1}^{\infty} a_n$ **converges to the sum** s if $\lim_{n \to \infty} s_n$ exists and equals s.

If the series does not converge it is said to **diverge**.

As for real series, the symbols $\sum_{n=1}^{\infty} a_k$ are used to denote the series itself and the sum if it exists.

Many of the properties of convergent series of real numbers have direct analogues for convergent series of complex numbers. We list some of them below, without proof. It is easy to construct the proofs if you know the corresponding proofs for real series. The only property of the absolute value of real numbers that is used is the triangle inequality ($|x + y| \leq |x| + |y|$): since the corresponding statement for the modulus of complex numbers is true, the proofs need very little alteration.

Results about Sequences and Series of Complex Numbers

1. If $\{a_n\}$ converges, its limit is unique: in other words, if $l = \lim_{n \to \infty} a_n$ and $l' = \lim_{n \to \infty} a_n$, then $l = l'$.

2. If $\sum_{n=1}^{\infty} a_n$ converges, its sum is unique: in other words, if $\sum_{n=1}^{\infty} a_n$ converges to the sum s, and also to the sum s', then $s = s'$.

3. If $\{a_n\}$ converges to l and $\{b_n\}$ converges to m, then $\{a_n + b_n\}$ converges to $l + m$; if z is any complex number then $\{za_n\}$ converges to zl.

4. If $\sum_{n=1}^{\infty} a_n$ converges to the sum s and $\sum_{n=1}^{\infty} b_n$ converges to the sum t, then $\sum_{n=1}^{\infty} (a_n + b_n)$ converges to the sum $s + t$; if z is any complex number then $\sum_{n=1}^{\infty} za_n$ converges to the sum zs.

108

5. If $\sum_{n=1}^{\infty} a_n$ converges, then, given any $\varepsilon > 0$, there is a natural number N such that

$$|a_{n+1} + a_{n+2} + \cdots + a_{n+k}| < \varepsilon \text{ whenever } n \geqslant N \text{ and } k \geqslant 1.$$

In particular, if $\sum_{n=1}^{\infty} a_n$ converges, then $\lim_{n \to \infty} a_n = 0$, but, as for real series, the converse is not true.

Result 5 is the analogue of one half of the Cauchy criterion for convergence of real series. It is important to know whether the Cauchy criterion holds in its entirety: to know, in other words, whether the condition of Result 5 is sufficient, as well as necessary, for the convergence of $\sum_{n=1}^{\infty} a_n$. You may remember that one of the consequences of the Cauchy criterion for real series is that every absolutely convergent series is convergent. If this result is true for complex series, it ought to be helpful because $\sum_{n=1}^{\infty} |a_n|$ will be a series of real non-negative terms and we already have a number of useful tests for classifying such series. So we first tackle the Cauchy criterion.

Theorem 2 (The Cauchy Criterion)

The series of complex numbers $\sum_{n=1}^{\infty} a_n$ converges, if and only if, for every $\varepsilon > 0$, there is a natural number N such that,

$$|a_{n+1} + a_{n+2} + \ldots + a_{n+k}| < \varepsilon, \text{ whenever } n \geqslant N \text{ and } k \geqslant 1.$$

Proof

We have already pointed out that the necessity of the condition is quite straight-forward to prove, so we deal only with its sufficiency.

Let $s_n = \sum_{r=1}^{n} a_r$. The convergence of the series $\sum_{r=1}^{\infty} a_r$. is equivalent to the convergence of the sequence $\{s_n\}$, and the condition of the theorem is that for every $\varepsilon > 0$, there is some N such that $|s_{n+k} - s_n| < \varepsilon$ whenever $n \geqslant N$ and $k \geqslant 1$. Let $s_n = x_n + iy_n$. Then the sequences of real numbers $\{x_n\}$ and $\{y_n\}$ are both Cauchy sequences because

$$|x_{n+k} - x_n| \leqslant |(x_{n+k} - x_n) + i(y_{n+k} - y_n)|$$

$$= |s_{n+k} - s_n|,$$

and, similarly,

$$|y_{n+k} - y_n| \leqslant |s_{n+k} - s_n|.$$

So $|x_{n+k} - x_n| < \varepsilon$ and $|y_{n+k} - y_n| < \varepsilon$ whenever $n \geqslant N$ and $k \geqslant 1$. Now every Cauchy sequence of real numbers converges. Let $x = \lim_{n \to \infty} x_n$ and $y = \lim_{n \to \infty} y_n$. Then by Theorem 1, $\{s_n\}$ has limit $s = x + iy$ and the series $\sum_{n=1}^{\infty} a_n$ converges to s. ■

We now consider absolute convergence.

Definition

The series of complex numbers $\sum_{n=1}^{\infty} a_n$ is **absolutely convergent** (or **converges absolutely**) if the series $\sum_{n=1}^{\infty} |a_n|$ converges.

This definition is superficially identical to the corresponding definition for real series: but remember that $\sum_{n=1}^{\infty} a_n$ is a complex series whereas $\sum_{n=1}^{\infty} |a_n|$ is a real series. However, as with real series, absolute convergence does imply convergence for complex series, and our next task is to prove this.

Theorem 3

If the series of complex numbers $\sum_{n=1}^{\infty} a_n$ is absolutely convergent, then it is convergent.

Proof

By hypothesis, $\sum_{n=1}^{\infty} |a_n|$ converges; so, by the Cauchy criterion for real series, given $\varepsilon > 0$, there is some natural number N such that

$$\big| |a_{n+1}| + |a_{n+2}| + \ldots + |a_{n+k}| \big| < \varepsilon \text{ whenever } n \geqslant N \text{ and } k \geqslant 1.$$

But

$$\big| |a_{n+1}| + |a_{n+2}| + \ldots + |a_{n+k}| \big| = |a_{n+1}| + |a_{n+2}| + \ldots + |a_{n+k}|$$

and

$$|a_{n+1} + a_{n+2} + \ldots + a_{n+k}| \leqslant |a_{n+1}| + |a_{n+2}| + \ldots + |a_{n+k}|;$$

so, whenever $n \geqslant N$ and $k \geqslant 1$,

$$|a_{n+1} + a_{n+2} + \ldots + a_{n+k}| < \varepsilon.$$

Thus $\sum_{n=1}^{\infty} a_n$ converges, by the Cauchy criterion for complex series. ■

As we mentioned earlier, absolute convergence is particularly useful because, if $a_n \neq 0$, $\sum_{n=1}^{\infty} |a_n|$ is a series of positive real numbers, to which we can apply the tests of real analysis. By far the most important of these for the purpose of this unit is the *ratio test* which we may state as follows:

if $\lim_{n \to \infty} \dfrac{|a_{n+1}|}{|a_n|} = k$ and $k < 1$ the series $\sum_{n=1}^{\infty} a_n$ converges absolutely,

whereas, if $k > 1$ the series diverges.

(When $k > 1$, the series diverges because there is a natural number N such that $\dfrac{|a_{n+1}|}{|a_n|} > 1$ whenever $n > N$ and so $|a_{n+1}| > |a_n|$ whenever $n > N$. Thus $\lim_{n \to \infty} a_n \neq 0$, and so the series is divergent.)

Examples

1. If z is any complex number such that $|z| < 1$, then the series $\sum_{n=0}^{\infty} |z|^n$ converges—it is just a geometric series. Thus $\sum_{n=0}^{\infty} z^n$ converges absolutely, and so converges, whenever $|z| < 1$.

2. If z is any complex number such that $\text{Re } z > 1$, then the series $\sum_{n=1}^{\infty} \dfrac{1}{n^z}$ converges, where n^z denotes $\exp(z \log n)$.

   We have

$$\left| \frac{1}{n^z} \right| = |\exp(z \log n)|^{-1} = e^{-\text{Re}(z \log n)} = \frac{1}{n^{\text{Re} z}}.$$

But $\sum_{n=1}^{\infty} \frac{1}{n^{\mathrm{Re}\,z}}$ converges since $\mathrm{Re}\,z > 1$, and so $\sum_{n=1}^{\infty} \frac{1}{n^z}$ converges absolutely, and thus converges. (You will meet this series in *Unit 11, Analytic Functions*, and *Unit 15, Number Theory*. It defines the Riemann zeta function.)

3. The series $\sum_{n=0}^{\infty} \frac{z^n}{n!}$ converges for any complex number z.

Provided $z \neq 0$, the ratio of successive terms of the series $\sum_{n=0}^{\infty} \left| \frac{z^n}{n!} \right|$ is

$$\left| \frac{z^{n+1}}{(n+1)!} \right| \cdot \left| \frac{n!}{z^n} \right| = \left| \frac{z}{n+1} \right|,$$

and $\lim_{n \to \infty} \left| \frac{z}{n+1} \right| = 0$. Thus the series $\sum_{n=0}^{\infty} \frac{z^n}{n!}$ converges absolutely when $z \neq 0$, and so converges. The series evidently converges when $z = 0$, too. We shall see later in this unit that this series can be used to define the exponential function.

Summary

In this section we have presented the basic definitions concerning complex sequences and series, namely: convergence of sequences, convergence and absolute convergence of series.

Many results about real sequences and series carry over to complex sequences and series. In particular, a series which is absolutely convergent is convergent.

We have also seen the usefulness of the ratio test applied to the sequence $\sum_{n=1}^{\infty} |a_n|$.

Self-Assessment Questions

1. Fill in the empty boxes in the following statements.

   (i) If $\lim_{n \to \infty} z_n = l$, then

   $$\lim_{n \to \infty} \mathrm{Re}\, z_n = \boxed{}, \text{ and } \lim_{n \to \infty} \mathrm{Im}\, z_n = \boxed{}.$$

   (ii) $\sum_{n=1}^{\infty} |a_n|$ converges $\boxed{}$ $\sum_{n=1}^{\infty} a_n$ converges.

   (Use one of $\Rightarrow$, $\Leftarrow$ and $\Leftrightarrow$.)

   (iii) The series $\sum_{n=0}^{\infty} \frac{z^n}{n!}$ converges for $z \in \boxed{}$.

   (iv) The series $\sum_{n=1}^{\infty} a_n$ converges $\boxed{}$ for every $\varepsilon > 0$, there is a natural number N such that

   $$|a_{n+1} + a_{n+2} + \ldots + a_{n+k}| < \varepsilon, \text{ whenever } n \geqslant N, k \geqslant 1.$$

   (Use one of $\Rightarrow$, $\Leftarrow$ and $\Leftrightarrow$.)

2. Write down the definition of convergence of a complex sequence.

Solutions

1. (i) Re l, Im l. (ii) $\Rightarrow$. (iii) **C**. (iv) $\Leftrightarrow$.

2. See page 107.

6.2 PROBLEMS

1. Prove Theorem 1 (page 108).

2. (a) Show that if the complex series $\sum_{n=1}^{\infty} a_n$ and $\sum_{n=1}^{\infty} b_n$ both converge absolutely then so does $\sum_{n=1}^{\infty} c_n$, where $c_n = a_n + b_n$. Show also that

$$\sum_{n=1}^{\infty} c_n = \sum_{n=1}^{\infty} a_n + \sum_{n=1}^{\infty} b_n.$$

   (b) (i) What can you say about the series $\sum_{n=1}^{\infty} \operatorname{Re} a_n$ and the series $\sum_{n=1}^{\infty} \operatorname{Im} a_n$ if the complex series $\sum_{n=1}^{\infty} a_n$ converges absolutely?

   (ii) What can you say about the complex series $\sum_{n=1}^{\infty} a_n$ if both the series $\sum_{n=1}^{\infty} \operatorname{Re} a_n$ and $\sum_{n=1}^{\infty} \operatorname{Im} a_n$ converge absolutely?

   (iii) Show that absolute convergence of a complex series implies convergence.

3. Show that if the complex series $\sum_{n=1}^{\infty} a_n$ converges absolutely, then $\sum_{n=1}^{\infty} |a_n| \geqslant \left| \sum_{n=1}^{\infty} a_n \right|$. (You may assume that if $\{s_n\}$ is any convergent sequence of real numbers and $s = \lim_{n \to \infty} s_n$, then $\lim_{n \to \infty} |s_n| = |s|$; and that if there is some real number t such that $s_n \leqslant t$ for all n, then $s \leqslant t$.)

4. In this problem you are asked to establish some results about the series obtained from a complex series $\sum_{n=1}^{\infty} a_n$ by throwing away the first $N - 1$ terms. We denote this series by $\sum_{n=N}^{\infty} a_n$.

   (a) Show that if $\sum_{n=1}^{\infty} a_n$ converges to the sum s then $\sum_{n=N}^{\infty} a_n$ converges to $s - \sum_{n=1}^{N-1} a_n$. In other words,

$$\sum_{n=N}^{\infty} a_n = \sum_{n=1}^{\infty} a_n - \sum_{n=1}^{N-1} a_n.$$

   (b) Show that if $\sum_{n=1}^{\infty} a_n$ converges absolutely, then so does $\sum_{n=N}^{\infty} a_n$.

   (c) Show that if $\sum_{n=1}^{\infty} a_n z^n$, where z is a fixed complex number, converges absolutely then

$$\left| \sum_{n=N}^{\infty} a_n z^n \right| \leqslant \sum_{n=N}^{\infty} |a_n| |z^n|.$$

Solutions

1. (i) Suppose $\lim_{n \to \infty} a_n = l$. Then, given $\varepsilon > 0$, there is a natural number N such that, if $n > N$, then

$$|a_n - l| < \varepsilon.$$

But $|\alpha_n - \alpha| \leqslant |a_n - l|$ and $|\beta_n - \beta| \leqslant |a_n - l|$.

Thus for $n > N$, $|\alpha_n - \alpha| < \varepsilon$ and $|\beta_n - \beta| < \varepsilon$, and so

$$\lim_{n \to \infty} \alpha_n = \alpha \text{ and } \lim_{n \to \infty} \beta_n = \beta.$$

(ii) Suppose $\lim_{n \to \infty} \alpha_n = \alpha$ and $\lim_{n \to \infty} \beta_n = \beta$. Then, given $\varepsilon > 0$, there is a natural number N such that, if $n > N$, then

$$|\alpha_n - \alpha| < \frac{\varepsilon}{2} \quad \text{and} \quad |\beta_n - \beta| < \frac{\varepsilon}{2}.$$

But

$$|a_n - l| \leqslant |\alpha_n - \alpha| + |\beta_n - \beta|$$
$$< \varepsilon, \text{ if } n > N,$$

and so $\lim_{n \to \infty} a_n = l$.

2. (a) We give a proof from first principles. The comparison test can also be used: see solution (b) (ii).

Given $\varepsilon > 0$, there is a natural number N such that, for $n \geqslant N$ and $k \geqslant 1$,

$$|a_{n+1}| + |a_{n+2}| + \ldots + |a_{n+k}| < \frac{\varepsilon}{2}$$

and

$$|b_{n+1}| + |b_{n+2}| + \ldots + |b_{n+k}| < \frac{\varepsilon}{2}.$$

Then, for $n \geqslant N$ and $k \geqslant 1$,

$$|c_{n+1}| + |c_{n+2}| + \ldots + |c_{n+k}|$$
$$= |a_{n+1} + b_{n+1}| + |a_{n+2} + b_{n+2}| + \ldots + |a_{n+k} + b_{n+k}|$$
$$\leqslant |a_{n+1}| + |a_{n+2}| + \ldots + |a_{n+k}|$$
$$+ |b_{n+1}| + |b_{n+2}| + \ldots + |b_{n+k}|$$
$$< \varepsilon,$$

and so $\sum_{n=1}^{\infty} c_n$ is absolutely convergent.

Now suppose that $\sum_{n=1}^{\infty} a_n = a$ and $\sum_{n=1}^{\infty} b_n = b$. Then, given $\varepsilon > 0$, there is a natural number N such that, for $n \geqslant N$,

$$\left| \sum_{k=1}^{n} a_k - a \right| < \frac{\varepsilon}{2} \quad \text{and} \quad \left| \sum_{k=1}^{n} b_k - b \right| < \frac{\varepsilon}{2}.$$

Thus, for $n \geqslant N$,

$$\left| \sum_{k=1}^{n} c_k - (a + b) \right| = \left| \sum_{k=1}^{n} a_k + \sum_{k=1}^{n} b_k - (a + b) \right|$$
$$\leqslant \left| \sum_{k=1}^{n} a_k - a \right| + \left| \sum_{k=1}^{n} b_k - b \right|$$
$$< \varepsilon,$$

and so $\sum_{n=1}^{\infty} c_n = a + b$.

(b) (i) We have $|\operatorname{Re} a_n| \leqslant |a_n|$ and $|\operatorname{Im} a_n| \leqslant |a_n|$.

Since $\sum_{n=1}^{\infty} |a_n|$ converges, the comparison test shows that $\sum_{n=1}^{\infty} \operatorname{Re} a_n$ and $\sum_{n=1}^{\infty} \operatorname{Im} a_n$ converge absolutely (and so converge).

(ii) If $\sum_{n=1}^{\infty} |\operatorname{Re} a_n|$ and $\sum_{n=1}^{\infty} |\operatorname{Im} a_n|$ converge, $\sum_{n=1}^{\infty} (|\operatorname{Re} a_n| + |\operatorname{Im} a_n|)$ converges. Since $|a_n| \leqslant |\operatorname{Re} a_n| + |\operatorname{Im} a_n|$, the comparison test tells us that $\sum_{n=1}^{\infty} a_n$ converges absolutely.

(iii) From (i) we deduce that if $\sum_{n=1}^{\infty} |a_n|$ converges then so do $\sum_{n=1}^{\infty} \operatorname{Re} a_n$ and $\sum_{n=1}^{\infty} \operatorname{Im} a_n$, and from Theorem 1 we deduce that $\sum_{n=1}^{\infty} a_n$ converges. So absolute convergence implies convergence.

3. We have $\lim_{N \to \infty} \sum_{n=1}^{N} a_n = \sum_{n=1}^{\infty} a_n$, and $\left| \sum_{n=1}^{\infty} a_n \right| = \lim_{N \to \infty} \left| \sum_{n=1}^{N} a_n \right|$.

   Further, $\left| \sum_{n=1}^{N} a_n \right| \leqslant \sum_{n=1}^{N} |a_n|$ for all N, and so $\lim_{N \to \infty} \left| \sum_{n=1}^{N} a_n \right| \leqslant \lim_{N \to \infty} \sum_{n=1}^{N} |a_n|$. Thus

   $$\left| \sum_{n=1}^{\infty} a_n \right| \leqslant \sum_{n=1}^{\infty} |a_n|.$$

4. (a) Given $\varepsilon > 0$, we can find a natural number N_0 such that, for $n \geqslant \max(N, N_0)$

   $$\left| \sum_{k=1}^{n} a_k - s \right| < \varepsilon.$$

   But

   $$\sum_{k=1}^{n} a_k = \sum_{k=1}^{N} a_k + \sum_{k=N+1}^{n} a_k,$$

   and so

   $$\left| \sum_{k=N+1}^{n} a_k - \left(s - \sum_{k=1}^{N} a_k \right) \right| < \varepsilon.$$

   Hence $\sum_{k=N+1}^{\infty} a_k$ converges to $s - \sum_{k=1}^{N} a_k$.

   (b) This follows from (a).

   (c) We have

   $$\left| \sum_{n=N}^{m} a_n z^n \right| \leqslant |a_N z^N| + |a_{N+1} z^{N+1}| + \ldots + |a_m z^m|$$

   $$= |a_N||z^N| + |a_{N+1}||z^{N+1}| + \ldots + |a_m||z^m|.$$

   Since $\sum_{n=1}^{\infty} a_n z^n$ is absolutely convergent, $\sum_{n=1}^{\infty} |a_n||z^n|$ converges, and so, by (b) $\sum_{n=N}^{\infty} |a_n||z^n|$ converges. It follows, therefore, that

   $$\lim_{m \to \infty} \left| \sum_{n=N}^{m} a_n z^n \right| \leqslant \lim_{m \to \infty} \sum_{n=N}^{m} |a_n||z^n|,$$

   that is

   $$\left| \sum_{n=N}^{\infty} a_n z^n \right| \leqslant \sum_{n=N}^{\infty} |a_n||z^n|.$$

6.3 COMPLEX POWER SERIES

A series of the form $\sum_{n=0}^{\infty} a_n(z - \alpha)^n = a_0 + a_1(z - \alpha) + a_2(z - \alpha)^2 + \ldots$ where the numbers $a_0, a_1, a_2, \ldots$ and α are (fixed) complex numbers is called a **power series centred at** α (or **about** α). (Note that, conventionally, $(z - \alpha)^0 = 1$.) We may define a function f by $f(z) = \sum_{n=0}^{\infty} a_n(z - \alpha)^n$ where the domain of f is the set of those z for which the series converges.

In this section we are going to investigate the properties of functions defined in this way. In particular, we shall discover that the domain of such a function—the set of points at which the series converges—is of a rather simple kind, and that the function is necessarily analytic on any open set in its domain.

$*$ $*$ $*$ $*$ $*$ $*$ $*$ $*$

To simplify the exposition slightly, we shall concentrate on power series of the form $\sum_{n=0}^{\infty} a_n z^n$, that is power series centred at 0. There are analogues of the results we are about to prove for power series centred at $\alpha \neq 0$, and you should be able to provide them for yourself.

You may recall that in *Unit M231 15, Uniform Convergence*, we showed that the set of points on which a real power series converges can be described as an interval, provided one is prepared to accept a single point, and the whole real line, as intervals. The situation for complex power series is much the same, except that we replace "interval" by "disc". To show that the sets on which complex power series converge are discs, we first show that if the series $\sum_{n=0}^{\infty} a_n z_0^n$ converges for some $z_0 \neq 0$, then $\sum_{n=0}^{\infty} a_n z^n$ converges whenever $|z| < |z_0|$. (If $z_0 = 0$, then clearly the series converges.) In fact, we shall see that it converges *absolutely* for $|z| < |z_0|$. It follows that the set $\{z : \sum_{n=0}^{\infty} a_n z^n \text{ converges}\}$ has the property that if it contains a non-zero z_0 then it contains the open disc $\{z : |z| < |z_0|\}$, and it is not difficult to convince oneself that a nonempty set with this property can be only a disc or the whole plane. (We shall prove it soon, but think about it for a minute or so before you read on.) First we prove the following lemma.

Lemma

If $\sum_{n=0}^{\infty} a_n z_0^n$ converges, and $z_0 \neq 0$, then $\sum_{n=0}^{\infty} a_n z^n$ converges absolutely for all z such that $|z| < |z_0|$.

Proof

Since $\sum_{n=0}^{\infty} a_n z_0^n$ converges, $\lim_{n \to \infty} |a_n z_0^n| = 0$. In particular, the sequence $\{|a_n z_0^n|\}$ must be bounded: that is, there is some number K such that $|a_n z_0^n| \leqslant K$ for all n. Then

$$|a_n z^n| = |a_n z_0^n| \cdot \left| \frac{z}{z_0} \right|^n \leqslant K \cdot \left| \frac{z}{z_0} \right|^n.$$

When $\left| \dfrac{z}{z_0} \right| < 1$, the real series $\sum_{n=0}^{\infty} K \cdot \left| \frac{z}{z_0} \right|^n$ converges, and so does $\sum_{n=0}^{\infty} |a_n z^n|$, by the comparison test. That is to say, $\sum_{n=0}^{\infty} a_n z^n$ is absolutely convergent, and thus convergent, when $|z| < |z_0|$. ∎

116

Theorem 4

Let $A = \{z : \sum_{n=0}^{\infty} a_n z^n \text{ converges}\}$. Then one of the following possibilities must hold:

(i) $A = \{0\}$;

(ii) there is $r > 0$ such that A consists of the disc $\{z : |z| < r\}$, possibly together with some or all of its boundary points;

(iii) $A = \mathbb{C}$.

Proof

If A is not $\{0\}$ or $\mathbb{C}$, then we can find z_1 and z_2 such that $|z_1| > 0$ and

$$\sum_{n=0}^{\infty} a_n z_1^n \text{ converges and } \sum_{n=0}^{\infty} a_n z_2^n \text{ diverges.}$$

From the lemma we know that $|z_1| \leqslant |z_2|$.

Let $r = \sup\{|z| : z \in A\}$. This exists because the set is nonempty (it contains $|z_1|$) and is bounded above (by $|z_2|$) for if there were some z in A such that $|z| > |z_2|$ then it would follow from the lemma that z_2 would also be in the set. Further $r > 0$ because $|z_1| > 0$.

Let z be such that $|z| < r$. There is a $z_0 \in A$ such that

$$|z| < |z_0| < r,$$

and so, by lemma, $\sum_{n=0}^{\infty} a_n z^n$ converges absolutely, and hence converges and so $z \in A$.

Thus, A contains the open disc $\{z : |z| < r\}$, and from the definition of r, A cannot contain any z for which $|z| > r$. We have proved nothing about points on the circle $|z| = r$, which may or may not belong to A. ∎

(Perhaps the easiest way to remember this proof is as follows. Call a circle centre the origin inside which the series converges a C-circle and one outside which it diverges a D-circle. Circles inside a C-circle must be C-circles and outside a D-circle must be D-circles. Each point of the plane must lie on one of these circles and there is only one circle common to both sets of circles.)

We have proved, then, that the set of points on which a power series converges is a disc (in a slightly generalised sense); the radius of this disc is called the **radius of convergence** of the power series. We have to allow that the radius of convergence may be 0 or ∞, to cover the cases $A = \{0\}$ and $A = \mathbb{C}$, respectively: the saving in effort in being able to cover all three cases in one expression is worthwhile.

If the radius of convergence of the series $\sum_{n=0}^{\infty} a_n z^n$ is finite and non-zero, and equal to R, the circle $\{z : |z| = R\}$ is called the **circle of convergence**. We have shown that, for $|z| < R$, the series not only converges—it converges absolutely. We have not discussed what happens *on* the circle of convergence, for the simple reason that anything may happen there. For example, each of the three series $\sum_{n=1}^{\infty} \frac{z^n}{n^2}$, $\sum_{n=1}^{\infty} \frac{z^n}{n}$ and $\sum_{n=1}^{\infty} z^n$ has radius of convergence equal to 1. The first converges everywhere on the circle of convergence $|z| = 1$ (in fact it converges absolutely); the second converges at $z = -1$ and diverges at $z = 1$ (that is, it converges at one point at least of its circle of convergence and diverges at one point at least); the third converges nowhere on its circle of convergence. (This last assertion follows from the fact that $\lim_{n \to \infty} z^n \neq 0$ if $|z| = 1$.)

117

One thing can be said about the behaviour of a series on its circle of convergence: if it converges *absolutely* at any one point on the circle, it converges absolutely at all points of the circle. The series $\sum_{n=1}^{\infty} \dfrac{z^n}{n^2}$ is an example. The reason for this is that the condition for absolute convergence of the series $\sum_{n=0}^{\infty} a_n z^n$ at any point z such that $|z| = R$ is that $\sum_{n=0}^{\infty} |a_n| R^n$ converges—and this condition is the same for all points of the circle.

We have seen, then, that if $f(z)$ is defined as the sum of a power series then it is defined for $z = 0$, or for all z, or for z in $\{z : |z| < R\} \cup B$, where B is some subset of $\{z : |z| = R\}$. We now turn our attention to the "sum function" f and investigate its properties. We prove first of all that f is continuous on a disc.

Theorem 5

Let $D = \{z : |z| < k\}$. If $f(z) = \sum_{n=0}^{\infty} a_n z^n$, $z \in D$, then f is continuous on D.

Proof

Consider the sequence $\{f_n(z)\}$ where $f_n(z) = \sum_{r=0}^{n} a_r z^r$ for $z \in D$. Each of the functions f_n is continuous on D. We have to prove that the function f where $f(z) = \lim_{n \to \infty} f_n(z)$ is also continuous.

Consider $z_0 \in D$ and $z_0 + h \in D$. We have to prove something about $|f(z_0 + h) - f(z_0)|$ and all we know is something about the relation between $f(z_0 + h)$ and the sequence $\{f_n(z_0 + h)\}$ and the relation between $f(z_0)$ and the sequence $\{f_n(z_0)\}$. So we shall have to rewrite $|f(z_0 + h) - f(z_0)|$ as follows:

$$|f(z_0 + h) - f(z_0)| = |f(z_0 + h) - f_N(z_0 + h) + f_N(z_0 + h)$$
$$- f_N(z_0) + f_N(z_0) - f(z_0)|$$
$$\leqslant |f(z_0 + h) - f_N(z_0 + h)|$$
$$+ |f_N(z_0 + h) - f_N(z_0)|$$
$$+ |f_N(z_0) - f(z_0)|.$$

We know that we can make the last expression as small as we please by choosing N large enough because $\{f_n(z_0)\}$ converges to $f(z_0)$. Similarly we can make the middle expression as small as we please by making h small enough (though how small will depend on N) because each function f_N is continuous. (Each f_N is a polynomial.) We can also make the first expression as small as we please by taking N large enough. It would seem then that we can make both the first and last expressions small just by choosing the larger of the two required N's. But that will not do because, when we start choosing h suitably to prove continuity, we might have to keep readjusting N and so get into a terrible mess. If we could find an N to make $|f_N(z) - f(z)|$ less than any given quantity for *all* $z \in D$ then once we make $|f(z_0 + h) - f_N(z_0 + h)|$ small enough, we are guaranteed that it will keep small as we vary h. We can not quite do that, but we can do something almost as good. We can make this remainder as small as we please for all z in any closed disc contained in D, as follows.

For any $z \in D$, $\{f_n(z)\}$ is convergent, so

$$|f(z) - f_N(z)| = \left| \sum_{n=N+1}^{\infty} a_n z^n \right|$$
$$\leqslant \sum_{n=N+1}^{\infty} |a_n||z^n|, \quad \text{by Problem 4(c) of Section 6.2,}$$
$$< \sum_{n=N+1}^{\infty} |a_n| k_1^n, \quad \text{if } z \in \{z : |z| < k_1, k_1 < k\}.$$

118

Since $\sum_{n=0}^{\infty} a_n k_1^n$ converges for all $k_1 < k$, given k_1 and $\varepsilon > 0$, we can find N such

that $\sum_{n=N}^{\infty} a_n k_1^n < \varepsilon$ and so $|f(z) - f_N(z)| < \varepsilon$ for *all* z in the disc $D_1 = \{z : |z| < k_1\}$.
With an eye on the proof that we are working on we might as well use $\varepsilon/3$
instead of ε. We now know that given any $\varepsilon > 0$ we can find N_0 such that

$$|f(z_0 + h) - f(z_0)| \leqslant \frac{\varepsilon}{3} + |f_N(z_0 + h) - f_N(z_0)| + \frac{\varepsilon}{3}$$

for $N > N_0$ and any $z_0 \in D_1$ and $z_0 + h \in D_1$.

Since f_N is continuous we can find $\delta > 0$ such that $z_0 + h \in D_1$ and

$$|f_N(z_0 + h) - f_N(z_0)| < \varepsilon/3$$

for all h in the disc $\{z : |z - z_0| < \delta\}$. Continuity of f at z_0 follows for any
$z_0 \in D_1$.

Now, given any $z_0 \in D$, k_1 can be chosen so that $z_0 \in D_1$ and so we deduce that
f is continuous on the whole of the open disc D.

(If it is obscure to you why we should have to work in the disc D_1, strictly
contained in D, see Problem 4 of Section 6.4. The point is that given k_1, we can
make $|f(z) - f_N(z)|$ less than any given $\varepsilon > 0$ for *all* z in D_1 just by choosing N
large enough. Problem 4 of Section 6.4 shows that this cannot be said for *all*
$z \in D$. But given any $z \in D$ we can find a suitable k_1 such that $z \in D_1$ and then
the argument of this proof goes through.) ■

Having established the continuity of f on D, we know that we have a function
that we can integrate along any contour Γ in D. The natural question to ask is

whether $\int_\Gamma f = \lim_{n \to \infty} \int_\Gamma f_n$.

So we propose and prove the following theorem.

Theorem 6

If $f_n(z) = \sum_{r=0}^{n} a_r z^r$, $z \in D = \{z : |z| < k\}$, and if $f(z) = \lim_{n \to \infty} f_n(z)$, then, for any

contour $\Gamma \subset D$, $\lim_{n \to \infty} \int_\Gamma f_n$ exists and equals $\int_\Gamma f$.

Proof

Theorem 5 assures us of the existence of $\int_\Gamma f$; also we have

$$\left| \int_\Gamma f - \int_\Gamma f_n \right| = \left| \int_\Gamma (f - f_n) \right|.$$

By the Estimation Theorem (see *Unit 4, Integration*), $\left| \int_\Gamma (f - f_n) \right|$ is bounded

above by $M_n L$, where L is the length of Γ and M_n an upper bound of $|f(z) - f_n(z)|$
on Γ; if we can choose M_n such that $\lim_{n \to \infty} M_n = 0$, we are done. We can do this
by finding such an upper bound for $|f(z) - f_n(z)|$, where z is not just on Γ but
anywhere in $D_1 = \{z : z \leqslant k_1\}$ where k_1 is chosen so that Γ lies in D_1: see Fig. 1.
(This can always be done because D is open.) This is exactly the bound we
obtained in the proof of the previous theorem. So, if we write

$$M_n = \sum_{r=n}^{\infty} |a_r| k_1^r,$$

we can choose N_0 large enough such that for any $\varepsilon > 0$,

$$M_n < \frac{\varepsilon}{L} \qquad \text{for all } n > N_0.$$

We therefore conclude that $\lim_{n \to \infty} \int_\Gamma f_n$ exists and is equal to $\int_\Gamma f$. ■

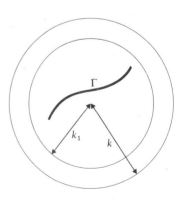

Fig. 1

In terms of power series, this theorem tells us that we can integrate term by term. In other words, if

$$f(z) = \sum_{r=0}^{\infty} a_r z^r, \quad z \in D,$$

then, for any contour Γ in D.

$$\int_{\Gamma} f(z)dz = \sum_{r=0}^{\infty} a_r \int_{\Gamma} z^r \, dz.$$

In fact, since $z \longrightarrow z^r$ is analytic on D, each of the integrals on the right is independent of the contour Γ, and so we can write

$$\int_{[0,z]} f = \sum_{r=0}^{\infty} \frac{a_r z^{r+1}}{r+1},$$

where we have taken Γ to be the line segment $[0, z]$.

So power series define continuous functions, which can be integrated by integrating term by term. What about differentiation? The fact that power series define analytic functions is an easy consequence of Morera's Theorem (see *Unit 5*).

Theorem 7

If $f(z) = \sum_{r=0}^{\infty} a_r z^r$ for all z in an open disc D, then f is analytic on D.

Proof

Let Δ be a triangle with vertices in D and let $f_n(z) = \sum_{r=0}^{n} a_r z^r$. Then $\int_{\Delta} f_n = 0$, because f_n is analytic on D for all n. Thus $\lim_{n \to \infty} \int f_n = 0$, and so $\int_{\Delta} f = 0$. We conclude that f is analytic on D, by Morera's Theorem. (See also Example 1 of Section 5.7 of *Unit 5*.) ∎

It is fairly clear that, having shown f to be analytic, our next step should be to see just what the derivative is— and we would be most surprised if it did not turn out that $f'(z) = \sum_{r=1}^{\infty} r a_r z^{r-1}$. We now prove this result.

Theorem 8

If $f(z) = \sum_{n=0}^{\infty} a_n z^n$, $z \in D = \{z : |z| < k\}$, then $f'(z) = \sum_{n=1}^{\infty} n a_n z^{n-1}$, $z \in D$.

Proof

Since D is open, for any $z \in D$ we can find k_1 such that $z \in D_1 = \{z : |z| < k_1\}$. Since f is analytic on D_1, if we write

$$f(z) = \sum_{n=0}^{N} a_n z^n + g_N(z), \quad z \in D_1,$$

then g_N is analytic on D_1, and so

$$f'(z) = \sum_{n=1}^{N} n a_n z^{n-1} + g_N'(z).$$

We clearly want to show that $\lim_{N \to \infty} g_N'(z) = 0$. Let C be the circle with centre z and radius ρ, where $\rho < k_1 - |z|$; then C lies in the disc D_1 (Fig. 2). Since g_N is analytic on D_1,

$$g_N'(z) = \frac{1}{2\pi i} \int_C \frac{g_N(w)}{(w-z)^2} \, dw, \quad \text{by Cauchy's Formula for the derivative.}$$

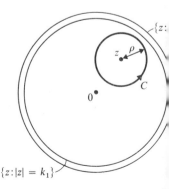

Fig. 2

120

Then, by the Estimation Theorem, $|g'_N(z)| \leqslant M_N \rho$ where M_N is an upper bound for $\left| \dfrac{g_N(w)}{(w-z)^2} \right|$, for $w \in C$; that is $\rho^2 M_N$ is an upper bound for $|g_N(w)|$, $w \in C$.

Now if $w \in D_1$,

$$|g_N(w)| \leqslant \sum_{n=N+1}^{\infty} |a_n w^n| \leqslant \sum_{n=N+1}^{\infty} |a_n| k_1^n.$$

Since $k_1 < k$, and the power series for f converges for $|z| < k$, we know that if $\rho^2 M_N = \sum_{n=N+1}^{\infty} |a_n| k_1^n$, then $\lim_{N \to \infty} M_N = 0$. Thus, since $|g'_n(z)| \leqslant M_N \rho$, we have $\lim_{N \to \infty} g'_N(z) = 0$, and the theorem is proved. ■

Summary

We have proved in this section that the set on which a power series converges is always a disc (or $\{0\}$ or C) whose boundary is the circle of convergence. Convergence at points on the circle of convergence has to be investigated independently. The radius of the circle of convergence is the radius of convergence of the power series. When it converges, a power series converges absolutely, except perhaps at points on the circle of convergence.

On any open disc on which a power series converges it defines a function that is analytic; the function may be integrated and differentiated by integrating or differentiating the series term by term.

Self-Assessment Questions

1. Fill in the empty boxes in the following statements.

    (i) Convergence of a power series on an open disc [ ] absolute convergence on that disc. (Use one of $\Rightarrow$, $\Leftarrow$ and $\Leftrightarrow$.)

    (ii) If $f(z) = \sum_{n=1}^{\infty} a_n z^n$ for all z in an open disc D, and $\Gamma \subset D$, then $\int_\Gamma f$

    exists because

    [   ].

    (iii) In the proof of Theorem 6 we were able to deduce that $\int_\Gamma f = 0$ from

    $\lim_{N \to \infty} \int f_N = 0$ because

    [   ].

    (iv) In the proof of Theorem 5 we could ensure that, for $N > N_0$, $|f(z_0 + h) - f_N(z_0 + h)|$ was small for all suitable h because

    [   ].

    (v) If $D = \{z : |z| < k\}$ and $z \in D$ then we can always find k_1 such that $0 \leqslant k_1 < k$ and $z \in \{z : |z| < k_1\}$ because

    [   ].

2. In Question 1(iv) what is meant by "suitable"?

121

Solutions

1. (i) $\Leftrightarrow$.

    (ii) f is continuous on D.

    (iii) $\displaystyle\int_\Gamma f = \lim_{N \to \infty} \int_\Gamma f_N$.

    (iv) for all z such that $|z| < k_1$, in particular for $z = z_0 + h$, $|f(z) - f_N(z)|$ is bounded by $\displaystyle\sum_{n=N}^{\infty} a_n k_1^n$ where $|z_0| < k_1 < k$.

    (v) D is open.

2. $z_0 + h \in D_1$.

6.4 PROBLEMS

1. Show that the function $f(z) = \sum_{n=0}^{\infty} \dfrac{z^n}{n!}$ is defined for all $z \in \mathbf{C}$ and that $f' = f$.

2. Show that the functions

$$f(z) = \sum_{n=0}^{\infty} \frac{(-1)^n z^{2n+1}}{(2n+1)!}, \quad g(z) = \sum_{n=0}^{\infty} \frac{(-1)^n z^{2n}}{(2n)!}$$

are defined for all $z \in \mathbf{C}$ and that $f' = g$ and $g' = -f$.

3. (a) Find the radius of convergence of each of the following series.

   (i) $\displaystyle\sum_{n=1}^{\infty} \frac{z^n}{n^3}$.

   (ii) $\displaystyle\sum_{n=1}^{\infty} nz^n$.

   (iii) $\displaystyle\sum_{n=1}^{\infty} \frac{z^n}{n^n}$.

   (iv) $\displaystyle\sum_{n=1}^{\infty} k^n(z-2)^n$, k fixed.

   (v) $\displaystyle\sum_{n=1}^{\infty} z^{2^n}$. (We shall meet this series later.)

   (b) Investigate the convergence of (i) and (v) in part (a), on the circle of convergence.

4. By the formula for the sum of the geometric series we know that

$$f_N(z) = \sum_{n=0}^{N-1} z^n = \frac{1 - z^N}{1 - z}, \quad z \neq 1,$$

and

$$f(z) = \sum_{n=0}^{\infty} z^n = \frac{1}{1 - z}, \quad |z| < 1.$$

Show that, for fixed k_1, given $\varepsilon > 0$, an N can be found such that

$$|f(z) - f_N(z)| < \frac{\varepsilon}{3}$$

for all z in $\{z : |z| \leqslant k_1 < 1\}$, but not for all z in $\{z : |z| < 1\}$.

Solutions

1. The ratio test gives convergence for all $z \in \mathbf{C}$ because

$$\lim_{n \to \infty} \left| \frac{z}{n+1} \right| = 0 \text{ for all } z \in \mathbf{C}.$$

$f' = f$ because differentiation term by term produces the same series. The function f is the exponential function, as you probably guessed, but we cannot prove that yet.

2. For f, the ratio test gives convergence for all $z \in \mathbf{C}$ because

$$\lim_{n \to \infty} \left| \frac{z^2}{(2n+3)(2n+2)} \right| = 0.$$

Similarly for g, because

$$\lim_{n \to \infty} \left| \frac{z^2}{(2n+2)(2n+1)} \right| = 0.$$

Differentiation term by term gives $f' = g$ and $g' = -f$. (The functions f and g are the sine and cosine functions but, as in Problem 1, we cannot prove that yet.)

3. (a) We use the ratio test.

   (i) $\displaystyle \lim_{n \to \infty} \left| \frac{z \cdot n^3}{(n+1)^3} \right| = |z|$, and so the radius of convergence is 1.

   (ii) $\displaystyle \lim_{n \to \infty} \left| \frac{(n+1)z}{n} \right| = |z|$, and so the radius of convergence is 1.

   (iii) $\displaystyle \lim_{n \to \infty} \left| \frac{z \cdot n^n}{(n+1)^{n+1}} \right| = 0$ for all z, and so the series converges for all z. The radius of convergence is ∞.

   (iv) $\displaystyle \lim_{n \to \infty} |k(z-2)| = |k||z-2|$, and so the series converges for $|z-2| < 1/|k|$. The radius of convergence is $1/|k|$.

   (v) $\displaystyle \lim_{n \to \infty} \left| \frac{z^{2^{n+1}}}{z^{2^n}} \right| = \lim_{n \to \infty} |z^{2^n}|$, and so the series converges for $|z| < 1$. The radius of convergence is 1.

   (b) In (i) of part (a), if $|z| = 1$, $\displaystyle \sum_{n=1}^{\infty} \left| \frac{z^n}{n^3} \right|$ becomes $\displaystyle \sum_{n=1}^{\infty} \frac{1}{n^3}$ and so we get absolute convergence on the circle of convergence. In (v) of part (a), if $|z| = 1$, $\displaystyle \sum_{n=1}^{\infty} z^{2^n}$ is divergent, because $\displaystyle \lim_{n \to \infty} z^{2^n} \neq 0$.

4. If $z \in \{z : |z| \leqslant k_1 < 1\}$, we have

$$|f(z) - f_N(z)| \leqslant \sum_{n=N}^{\infty} k_1^n = \frac{k_1^N}{1 - k_1} < \frac{\varepsilon}{3}$$

for sufficiently large N (depending on k_1).

If $z \in \{z : |z| < 1\}$, we have

$$|f(z) - f_N(z)| = \frac{|z|^N}{|1 - z|}.$$

Choose $z = k_1 + \delta$, where $0 < \delta < 1 - k_1$. Then

$$|f(z) - f_N(z)| = \frac{|k_1 + \delta|^N}{1 - k_1 - \delta} > \frac{k_1^N}{1 - k_1}.$$

So, if $\displaystyle \frac{k_1^N}{1 - k_1} < \frac{\varepsilon}{3} < \frac{|k_1 + \delta|^N}{1 - k_1 - \delta}$, we have

$$|f(z) - f_N(z)| \not< \frac{\varepsilon}{3}.$$

6.5 TAYLOR SERIES

We have seen that power series can be used to define analytic functions: we are now going to show that any analytic function can be represented by power series. The series are called the Taylor series of the function: a name which should be familiar to you from real analysis. We shall again see a marked contrast between real and complex analysis. Given *any* analytic complex function f and a point z in its domain, we can find a Taylor series which converges to $f(z)$: but, given a real function which is infinitely differentiable, although one can calculate a Taylor series, there is no guarantee that the series converges. For example, for the function $f(x) = 1/(1 + x^2)$, the Taylor polynomial of f at 0 does not converge at all if $|x| > 1$. See **Spivak**, Chapter 23.

The work in this section has important theoretical consequences which we shall develop in Section 6.7, but it also helps with a practical issue. From our work on power series it seems likely that we can approximate a function using a finite number of terms of a power series. If we are to do this then it would be useful to have some indication of the accuracy of this approximation; in other words we require a knowledge of a remainder term.

As in real analysis we are able to find a remainder term, which is given by an integral. Unlike the real analysis case, we do not have to resort to special cases to investigate the remainder—we can show that it can be made as small as we please for *any* function at any point in a disc on which that function is analytic. Furthermore, we are able to show that if a function is defined by a power series then that power series is indeed a Taylor series for the function.

One final point before we get down to work: we have seen that power series converge on discs, but since functions can be defined on all sorts of regions, it is clear that we cannot necessarily represent a function by one Taylor series throughout its domain. So first of all we should concentrate on functions defined on discs and derive some local results. Then, when in the hypotheses of a theorem we say that a function is analytic on a disc, this is to be regarded as a minimum requirement; it does not exclude the possibility that the function is analytic on a larger region.

$$* \qquad * \qquad * \qquad * \qquad * \qquad * \qquad * \qquad *$$

In Section 6.3 we showed that every power series with a finite non-zero radius of convergence k defines a function analytic on the open disc $D = \{z : |z| < k\}$. Furthermore, we have seen that if

$$f(z) = \sum_{n=0}^{\infty} a_n z^n, \quad z \in D,$$

then

$$f'(z) = \sum_{n=1}^{\infty} n a_n z^{n-1}, \quad z \in D.$$

Since f is analytic on D we know, from *Unit 5*, that it is infinitely differentiable on D. So $f^{(k)}$ is analytic on D and, applying the results of Section 6.3, we get

$$f''(z) = \sum_{n=2}^{\infty} n(n-1) a_n z^{n-2},$$

$$f'''(z) = \sum_{n=3}^{\infty} n(n-1)(n-2) a_n z^{n-3},$$

and so on.

125

In particular, we see that the values of the function and its derivatives at 0 are given by

$$f(0) = a_0, \quad f'(0) = a_1, \ldots, f^{(n)}(0) = n! a_n, \ldots \, .$$

If you are thinking along the right lines, you should be anticipating the next move. In Section 6.3 the series defined the function; and in particular, the coefficients specified $\dfrac{f^{(n)}(0)}{n!}$. Suppose we turn everything back to front. Suppose we know f; can we use $f(0), f'(0), f''(0)$, and so on, to calculate the coefficients, and so specify a power series that represents f? We clearly want to propose a theorem something like this:

If f is analytic on the disc $D = \{z : |z| < k\}$, then $f(z) = \sum_{n=0}^{\infty} \dfrac{f^{(n)}(0)}{n!} z^n$ for all z in some disc in D.

We have not labelled this as a "Theorem" because, as yet, we have no idea where this proposed series will converge. We know from Section 6.3 that if it converges at all for non-zero z then it will converge on a disc. We should like the series to converge on D itself, but, bearing in mind the things that happen with Taylor series in the real variable case, it would seem that we would be pushing our luck to expect this. (For example, the function $f(x) = \arctan x$ is continuous and differentiable for all real x, but its Taylor series centred at 0 converges only if $|x| \leqslant 1$. See Chapter 19 of **Spivak**.) We are even hoping for a lot in proposing that the series will converge to $f(z)$ even at all those points where it does converge! Notice, however, that we have chosen D to be an *open* disc; we already know from our work on power series that we are unlikely to be able to say anything at all about convergence on the boundary in the general case.

We shall set about a proof of our proposed theorem and see how things work out.

Let C be a circle in D, say $C = \{z : |z| = \rho\}$, $\rho < k$ (Fig. 3). Then by Cauchy's Formula,

$$f(z) = \frac{1}{2\pi i} \int_C \frac{f(w)}{w - z} \, dw, \quad \text{if } |z| < \rho.$$

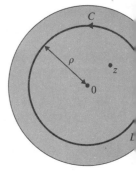

Fig. 3

Now

$$\frac{1}{w - z} = \frac{1}{w} \cdot \frac{1}{1 - \dfrac{z}{w}}.$$

The term $1 - \dfrac{z}{w}$ reminds us of the sum of a finite geometric series, because

$$1 + \frac{z}{w} + \left(\frac{z}{w}\right)^2 + \ldots + \left(\frac{z}{w}\right)^{n-1} = \frac{1 - \left(\dfrac{z}{w}\right)^n}{1 - \dfrac{z}{w}}, \quad z \neq w.$$

Hence

$$\frac{1}{1 - \dfrac{z}{w}} = 1 + \frac{z}{w} + \left(\frac{z}{w}\right)^2 + \ldots + \left(\frac{z}{w}\right)^{n-1} + \frac{\left(\dfrac{z}{w}\right)^n}{1 - \dfrac{z}{w}},$$

126

and so

$$\frac{1}{w-z} = \frac{1}{w} \cdot \frac{1}{1-\dfrac{z}{w}}$$

$$= \frac{1}{w}\left(1 + \frac{z}{w} + \left(\frac{z}{w}\right)^2 + \ldots + \left(\frac{z}{w}\right)^{n-1} + \frac{\left(\dfrac{z}{w}\right)^n}{1-\dfrac{z}{w}}\right)$$

$$= \frac{1}{w} + \frac{z}{w^2} + \frac{z^2}{w^3} + \ldots + \frac{z^{n-1}}{w^n} + \frac{z^n}{w^n} \cdot \frac{1}{w-z}.$$

Therefore,

$$f(z) = \frac{1}{2\pi i}\int_C \frac{f(w)}{w}\,dw + \frac{z}{2\pi i}\int_C \frac{f(w)}{w^2}\,dw + \ldots + \frac{z^{n-1}}{2\pi i}\int_C \frac{f(w)}{w^n}\,dw + z^n g_n(z),$$

where $g_n(z) = \dfrac{1}{2\pi i}\displaystyle\int_C \frac{f(w)}{w^n(w-z)}\,dw.$

But $\dfrac{1}{2\pi i}\displaystyle\int_C \frac{f(w)}{w^{k+1}}\,dw = \frac{f^{(k)}(0)}{k!}$, by Cauchy's Formula for derivatives.

So

$$f(z) = f(0) + \frac{f'(0)}{1!}z + \ldots + \frac{f^{(n-1)}(0)}{(n-1)!}z^{n-1} + z^n g_n(z).$$

We must now show that $|z^n g_n(z)|$ is small if n is large. Now by Theorem 12 of *Unit 2, Continuous Functions* (a function continuous on a closed bounded set is bounded on that set), f is certainly bounded on C because C is closed and bounded and f is continuous on C. Suppose, then, that $|f(w)| < M$, say, for $w \in C$. Moreover, $|w - z| \geqslant |w| - |z| = \rho - |z|$. Thus, by the Estimation Theorem,

$$|z^n g_n(z)| \leqslant \frac{|z|^n}{2\pi} \cdot \frac{M}{\rho^n(\rho - |z|)} \cdot 2\pi\rho$$

$$= \frac{M}{\rho - |z|} \cdot \left(\frac{|z|}{\rho}\right)^n \cdot \rho.$$

But $\dfrac{|z|}{\rho} < 1$, and so $\displaystyle\lim_{n\to\infty}\left(\frac{|z|}{\rho}\right)^n = 0$. Thus $|z^n g_n(z)|$ may be made as small as desired by choosing n sufficiently large.

This proves that f is represented by the series $\displaystyle\sum_{n=0}^{\infty} \frac{f^{(n)}(0)}{n!}z^n$ on $\{z : |z| < \rho\}$. Furthermore, we can say this for all values of ρ provided $0 < \rho < k$ and for any z for which $|z| < \rho$. But for *any* $z \in D$ we can *choose* ρ such that $|z| < \rho < k$ and hence the function is represented by the series throughout D. We therefore have proved the following theorem.

Theorem 9

If f is analytic on an open disc $D = \{z : |z| < k\}$, then the series

$$\sum_{n=0}^{\infty} \frac{f^{(n)}(0)}{n!}z^n$$

converges to $f(z)$ for all z in D.

Definition

The series $\displaystyle\sum_{n=0}^{\infty} \frac{f^n(0)}{n!}z^n$ is the **Taylor series for** f **at** 0.

127

Corollary

If $f(z) = \sum_{n=0}^{\infty} a_n z^n$ for all z in an open disc, then $a_n = \dfrac{f^{(n)}(0)}{n!}$ and the series is the Taylor series for f at 0.

Proof

Apply Theorem 8 to differentiate n times and put $z = 0$. ■

It is sometimes useful to retain the remainder term and express $f(z)$ as a finite series.

Theorem 10

If f is analytic on an open disc $D = \{z : |z| < k\}$, then

$$f(z) = \sum_{n=0}^{N-1} \frac{f^{(n)}(0)}{n!} z^n + z^N g_N(z), \quad z \in D,$$

where

$$g_N(z) = \frac{1}{2\pi i} \int_C \frac{f(w)}{w^N(w-z)} dw,$$

and C is any circle contained in D and enclosing z and 0. Furthermore, the function g_N is analytic on D and $g_N(0) = \dfrac{f^{(N)}(0)}{N!}$.

Proof

The only new item to prove here is the last sentence. We have

$$g_N(z) = \frac{1}{2\pi i} \int_C \frac{f(w)}{w^N(w-z)} dw, \quad z \in C.$$

Since the circle C is contained in D and does not contain 0, the function $w \longrightarrow f(w)/w^n$ is continuous on C. Thus, by Problem 6, Section 5.6 of *Unit 5*, g_N is analytic on the complement of C. It is therefore certainly analytic on the open disc bounded by C. For any $z \in D$, a suitable circle C can be found enclosing z and hence g_N is analytic on D.

We also have

$$g_N(0) = \frac{1}{2\pi i} \int_C \frac{f(w)}{w^{N+1}} dz$$

$$= \frac{f^N(0)}{N!}, \quad \text{by Cauchy's Formula for the } N\text{th derivative.} \quad ■$$

We remarked in Section 6.1 that we would be concentrating on power series of the form $\sum_{n=0}^{\infty} a_n z^n$. Extensions to series of the form $\sum_{n=0}^{\infty} a_n(z - \alpha)^n$ are in the main fairly obvious and the new proofs amount to little more than replacing z by $z - \alpha$. That was completely acceptable when our concern was power series as such, but we have now transferred our attention to functions which we are trying to represent as power series. Now Theorem 9 is of little use for functions whose domain does not contain the origin! The adjustment is obvious, but worth noting.

Theorem 11 (Taylor's Theorem)*

If f is analytic on an open disc $D = \{z : |z - \alpha| < k\}$, then the series

$$\sum_{n=0}^{\infty} \frac{f^{(n)}(\alpha)}{n!}(z - \alpha)^n$$

converges to $f(z)$ for all z in D.

Definition

The series $\sum_{n=0}^{\infty} \frac{f^{(n)}(\alpha)}{n!}(z - \alpha)^n$ is the **Taylor series for f at α**. The coefficients $\frac{f^{(n)}(\alpha)}{n!}$ are called the **Taylor coefficients**.

Theorem 10 also has a more general form.

Theorem 12

If f is analytic on an open disc $D = \{z : |z - \alpha| < k\}$, then

$$f(z) = \sum_{n=0}^{N-1} \frac{f^{(n)}(\alpha)}{n!}(z - \alpha)^n + (z - \alpha)^N g_N(z), \quad z \in D,$$

where

$$g_N(z) = \frac{1}{2\pi i}\int_C \frac{f(w)}{(w - \alpha)^N(w - z)}\,dw,$$

and C is any circle contained in D enclosing z and α. Furthermore, the function g_N is analytic on D and $g_N(\alpha) = \dfrac{f^{(N)}(\alpha)}{N!}$.

The finite series $\sum_{n=0}^{N-1} \frac{f^n(\alpha)}{n!}(z - \alpha)^n$ is referred to as the *Taylor approximation to f at α* (it is a polynomial in $(z - \alpha)$ of degree $N - 1$). The term $g_N(z) = (z - \alpha)^N g_N(z)$ is called the *remainder term*.

The proofs of Theorems 11 and 12 both follow along the lines of those for Theorems 9 and 10 by a "shift of origin" using the analytic function $z \longrightarrow z + \alpha$.

You may have noticed that once again we have an example of the power of the notion of analyticity. We have seen in *Unit 5* that the existence of the first derivative is sufficient to guarantee the existence (and the continuity) of all derivatives. We now see that it guarantees the existence and convergence of the Taylor series in any open disc on which the function is analytic.

We remarked at the beginning of this section that given *any* z in a region on which a function is analytic we could find a Taylor series that converges to $f(z)$. All we have proved so far is that if f is analytic on an open disc then we can represent it by a Taylor series throughout that disc. So, if f is analytic on a region R which is not a disc, we have no guarantee that, given a particular $\alpha \in R$, a Taylor series for f at α will give a representation for $f(z)$ for *every* z in R, because we may not be able to make the disc large enough to contain z but keep within R (Fig. 4).

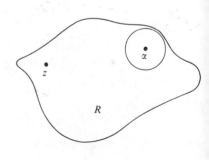

Fig. 4

---

* *Brook Taylor* (1685–1731) was born in Canada, but came to England and studied in Cambridge. Between the years 1712 and 1719, he wrote numerous papers on such subjects as oscillations, capillarity and projectiles. Taylor's Theorem (for a real variable) was discovered in 1715, but his proof contains no discussion of the remainder term or of convergence, and would certainly not be accepted today. The complex version of Taylor's Theorem given above is due to Cauchy.

Of course, the immediate problem is not severe because we can always find *another* circle enclosing z and then form the Taylor series inside this circle. In other words, we may well need more than one Taylor series to cover the whole region R.

We can also shed some interesting light on *real* power series. It is easy enough to show that for real x,

$$1 - x^2 + x^4 - x^6 + \ldots$$

converges to $1/(1 + x^2)$ for $|x| < 1$. But the methods of real analysis do not enable us to attach any significance to ± 1 in terms of the function $f(x) = 1/(1 + x^2)$ itself. However, the *complex* function $f(z) = 1/(1 + z^2)$ is not defined at $\pm i$ and cannot be assigned special values at $\pm i$ so as to make it analytic on any neighbourhood of either of these points (because it is not bounded on any punctured neighbourhood of either point), although it is analytic on every region not containing $\pm i$. Thus no disc with centre the origin in which the Taylor series converges can include $\pm i$ and thus the unit circle is the circle of convergence. This unit circle cuts out an interval on the real axis on which the real series converges (Fig. 5).

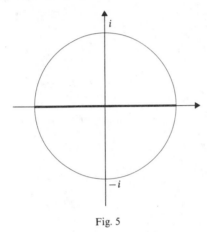

Fig. 5

In much the same way we can conclude at once that the binomial series for $(1 + x)^\alpha$ converges to $(1 + x)^\alpha$ for x in $(-1, 1)$ and that a Taylor series centre 0 for $\exp(-1/x^2)$ does not converge to $\exp(-1/x^2)$. (The function $f(z) = \exp(-1/z^2)$ is not analytic on any neighbourhood of 0, because it is unbounded on any punctured neighbourhood of 0, and hence not continuous on such a neighbourhood, and therefore not analytic.)

There is one result of a more global nature that we can prove before concluding this section; it concerns the remainder term $(z - \alpha)^N g_N(z)$. We have seen that in any disc centre α on which f is analytic, g_N is analytic and, of course, the sequence $\{g_N(z)\}$ has limit zero. Now even if the region R on which f is analytic is such that the positions of z and α make it impossible to include z in a disc centre α contained in R, we can still calculate the Taylor approximation to $f(z)$ at α. In such a case we clearly cannot say anything in general about the limit of $\{g_N(z)\}$ but we can prove that for any N, g_N is analytic on a neighbourhood of z. In fact, we can prove the following theorem.

Theorem 13

Let f be analytic on a region R. Then for any $\alpha \in R$ we can find a sequence $\{g_N(z)\}$ where each function g_N is analytic on R, such that for all $z \in R$,

$$f(z) = \sum_{n=0}^{N-1} \frac{f^{(n)}(\alpha)}{n!}(z - \alpha)^n + (z - \alpha)^N g_N(z)$$

and

$$g_N(\alpha) = \frac{f^{(N)}(\alpha)}{N!}.$$

130

Proof

Let $\alpha \in R$. Define g_N by

$$g_N(z) = \begin{cases} (z - \alpha)^{-N}\left(f(z) - \displaystyle\sum_{n=0}^{N-1} \frac{f^{(n)}(\alpha)}{n!}(z - \alpha)^n\right), & z \neq \alpha \\ \dfrac{f^{(N)}(\alpha)}{N!}, & z = \alpha. \end{cases}$$

It is clear that g_N is analytic everywhere on R except possibly on a neighbourhood of α. But given $\alpha \in R$ we can find an open disc centre α and contained in R. By the local form of this theorem, Theorem 12, g_N is analytic on this open disc. Thus g_N is analytic throughout R. ∎

Summary

The formula for the coefficients in the Taylor series of a complex function is the same as for real functions, but we can guarantee that the series will converge to the function throughout any open disc on which the function is analytic. The remainder term is given by an integral and does itself define an analytic function throughout any region on which the original function is analytic.

Self-Assessment Questions

1. Classify the following statements as True or False.

   (i) Any function can be represented by its Taylor series at α throughout its domain.

   (ii) The proof of the Corollary to Theorem 9 is valid because the remainder term is analytic.

2. Write down the first three terms in the Taylor series for f at α for each of the following functions f and points α.

   (i) $f(z) = \dfrac{1}{z}$; $\alpha = 1$.

   (ii) $f(z) = \sin z$; $\alpha = 0$.

   (iii) $f(z) = e^{1+z}$; $\alpha = 2$.

3. Write down the radius of convergence of each of the Taylor series in Question 2.

4. Given that

   $$f(z) = \frac{1}{1 + z^2} = 1 - z^2 + z^4 - \ldots, \quad |z| < 1,$$

   write down the value of $I = \displaystyle\int_C \frac{1}{w^k(1 + w^2)}dw$, where $C = \{z : |z| = \tfrac{1}{2}\}$, in terms of f.

Solutions

1. (i) False, for two reasons. Firstly, the function must be analytic on a neighbourhood of every point where it is represented by a Taylor series. Secondly, if the domain is not a disc, we cannot necessarily guarantee that the circle of convergence of the Taylor series can be made large enough to enclose every point in the domain.

   (ii) False: it is valid because a power series defines a function analytic on any open set on which it converges.

2. (i) $\dfrac{1}{z} = 1 - (z - 1) + (z - 1)^2 + \ldots$

   (ii) $\sin z = z - \dfrac{z^3}{3!} + \dfrac{z^5}{5!} - \ldots$

   (iii) $e^{1+z} = e^3 \left(1 + (z - 2) + \dfrac{(z - 2)^2}{z!} + \ldots \right).$

Each of these results can be obtained directly, but the first and third can also be obtained by a little trickery. For example, if $w = z - 1$, then $\dfrac{1}{z} = \dfrac{1}{1 + w}$. If one is allowed to assume the Taylor series for $\dfrac{1}{1 + w}$ at 0, the series for $\dfrac{1}{z}$ at 1 follows by substitution. In the third case, substitution of $w = z - 2$ gives

$$e^{1+z} = e^{3+w} = e^3 e^w = e^3 \left(1 + w + \frac{w^2}{2!} + \ldots \right),$$

and so on.

3. (i) 1, because the function ceases to be analytic at $z = 0$.

   (ii) ∞, because the function is entire.

   (iii) ∞, because the function is entire.

4. The integral I is $f^{(k-1)}(0) \cdot \dfrac{2\pi i}{(k - 1)!}$, by Cauchy's Formula for derivatives. From the given series, we see that $\dfrac{f^{(k-1)}(0)}{(k - 1)!} = 0$ if k is even and $(-1)^{(k-1)/2}$ if k is odd.

6.6 PROBLEMS

Some Remarks on the Practical Calculation of Taylor Series

The main emphasis so far in this unit has been on the development of the analytic ideas associated with Taylor series. The only explicit method that we have seen for calculating the coefficients is by successive differentiation. But, of course, just as in real analysis, there are short cuts that can be helpful in practical situations. As often happens in analysis we can use the methods of combining functions—addition, multiplication, division and composition—to extend results for elementary functions to more complicated ones. Thus, for example, you will see in Problem 3 below that the Taylor series for functions f and g can be multiplied to give the Taylor series for fg.

We also know that power series can be differentiated and integrated term by term. Since

$$\frac{1}{1+z} = 1 - z + z^2 - z^3 + \ldots, \quad |z| < 1,$$

we can deduce, by differentiation, that

$$\frac{-1}{(1+z)^2} = -1 + 2z - 3z^2 + \ldots, \quad |z| < 1,$$

and by multiplication by -1 (the Taylor Series for $z \longrightarrow -1$) that

$$\frac{1}{(1+z)^2} = 1 - 2z + 3z^2 + \ldots, \quad |z| < 1.$$

We could also deduce this result by direct multiplication:

$$(1 - z + z^2 - z^3 + z^4 - \ldots)(1 - z + z^2 - z^3 + z^4 - \ldots)$$
$$= 1 + z[-1 + (-1)] + z^2[1 + (-1)(-1) + 1] + \ldots$$
$$= 1 - 2z + 3z^2 + \cdots.$$

In Problem 4 below, the Taylor series of $1/f$ is discussed in terms of that of f. From this one can of course deduce techniques for finding the Taylor series of quotients of functions.

Composition of functions is another way of extending the catalogue of Taylor series. Since, for example

$$\frac{1}{1-z} = 1 + z + z^2 + \cdots, \quad |z| < 1,$$

and

$$\sin z = z - \frac{z^3}{3!} + \frac{z^5}{5!} \cdots,$$

we have

$$\sin \frac{1}{1-z} = (1 + z + z^2 + \cdots)$$
$$- \frac{(1 + z + z^2 + \cdots)^3}{3!}$$
$$+ \frac{(1 + z + z^2 + \cdots)^5}{5!}$$
$$+ \cdots$$
$$= (1 + z + z^2 + \cdots)$$
$$- \tfrac{1}{6}(1 + 3z + 6z^2 + \cdots)$$
$$+ \tfrac{1}{120}(1 + 5z + 15z^2 + \cdots)$$
$$+ \cdots.$$

Of course, it is highly likely that techniques such as these will not produce a formula for the general coefficient in a Taylor series, but often it is only the first few terms that are required. (In the last example above, we did not even get that.) The disc of convergence can often be predicted from the function f itself— we know that the Taylor series for f at α will converge in any disc with centre α on which f is analytic.

The only snag that might have occurred to you is that these tricks might produce the "wrong" Taylor series. How can we be sure that we have not got a Taylor series for some other function? Consider, for example $f(h(z))$. Let

$$f(z) = a_0 + a_1 z + \cdots + a_n z^n + g_n(z)$$
$$= p_n(z) + g_n(z),$$

and

$$h(z) = b_0 + b_1 z + \cdots + b_n z^n + k_n(z)$$
$$= q_n(z) + k_n(z),$$

where $g_n(z)$ and $k_n(z)$ each have the form

$$z^{n+1}(c_{n+1} + c_{n+2} z + \dots).$$

Thus

$$f(h(z)) = p_n(q_n(z) + k_n(z)) + \text{powers of } z \text{ higher than } z^n$$
$$= p_n(q_n(z)) + \text{powers of } z \text{ higher than } z^n.$$

So the first $n + 1$ terms in the series obtained in this way by substitution must be the terms of degree at most n in $p_n(q_n(z))$, and these are the first $n + 1$ terms of the required Taylor series. If they are the first terms of another Taylor series, the difference is only in subsequent terms.

---

1. Find the Taylor series at α for the following functions f and points α, and find the radius of convergence in each case. (It would be a good idea in this problem to use first principles, that is to say, differentiation, to find the coefficients.)

    (i) $f(z) = \dfrac{1}{1 + z}$; $\alpha = 0$.

    (ii) $f(z) = \dfrac{1}{1 + z}$; $\alpha = 1$.

    (iii) $f(z) = \dfrac{1}{z}$; $\alpha = 1$.

    (iv) $f(z) = \dfrac{1}{(1 - z)^2}$; $\alpha = 0$.

    (v) $f(z) = \dfrac{1}{(1 - z)^2}$; $\alpha = -1$.

    (vi) $f(z) = e^z$; $\alpha = 0$.

    (vii) $f(z) = \exp(k \operatorname{Log}(1 + z))$, k fixed; $\alpha = 0$.

2. Solve the differential equation $f''(z) + f(z) = 0$, by assuming that a solution of the form $f(z) = \sum\limits_{n=0}^{\infty} a_n z^n$ exists, and determining the coefficients a_n.

3. Suppose that $f(z) = \sum_{n=0}^{\infty} a_n z^n$ and $g(z) = \sum_{n=0}^{\infty} b_n z^n$ for all z in an open disc D

and that $f_N(z) = \sum_{n=0}^{N} a_n z^n$ and $g_N(z) = \sum_{n=0}^{N} b_n z^n$.

(i) Write down the coefficient of z^n in $f_N(z)g_N(z)$, where $n \leqslant N$.

(ii) If $c_n = \sum_{k=0}^{n} a_k b_{n-k}$, show, by considering the Taylor series of fg, that

the series $\sum_{n=0}^{\infty} c_n z^n$ converges to $f(z)g(z)$ at each point z in D.

(iii) Find the first four terms in the Taylor series at 0 for the function

$f(z) = \dfrac{e^z}{(1-z)^2}$, and give its radius of convergence.

4. (i) Suppose $f(z) = \sum_{n=0}^{\infty} a_n z^n$ for all z in some disc D and $a_0 \neq 0$. Show that

if $g = \dfrac{1}{f}$ and $g(z) = \sum_{n=0}^{\infty} b_n z^n$, then

$a_0 b_0 = 1,$

$a_1 b_0 + a_0 b_1 = 0,$

$a_2 b_0 + a_1 b_1 + a_0 b_2 = 0,$

$$\vdots$$

$\sum_{k=0}^{n} a_k b_{n-k} = 0.$

(ii) Use (i) to calculate the Taylor series at 0 for the function

$f(z) = \dfrac{1}{1+z}.$

(iii) Calculate the first three terms of the Taylor series at 0, of the functions

$z \longrightarrow \dfrac{1}{\cos z}$ and tan.

5. (i) If $f(z) = \dfrac{1}{1-z}$, $z \neq 1$, show that the Taylor series at α for f is

$$\frac{1}{1-\alpha} \sum_{k=0}^{\infty} \left(\frac{z-\alpha}{1-\alpha}\right)^k.$$

(ii) Find the radius of convergence of this series.

(iii) By substituting $w = \dfrac{z-\alpha}{1-\alpha}$, verify directly that at its points of convergence, the sum is $f(z)$.

(iv) Deduce directly that $f(z)$ is analytic on $\mathbf{C} - \{1\}$.

6. Let $f(z) = \sum_{n=0}^{\infty} a_n z^n$ for all z in $\{z : |z| < k\}$ and let

$$M(r) = \sup\{|f(z)| : |z| = r < k\}.$$

(i) Show that

$$|a_n| \leqslant \frac{M(r)}{r^n}.$$

If you are short of time and have not studied M201, Linear Mathematics, or MST 282, Mechanics and Applied Calculus, then you should omit parts (ii), (iii) and (iv).

(ii) Show that

$$a_n = \frac{1}{2\pi r^n} \int_0^{2\pi} f(re^{i\theta}) e^{-in\theta} \, d\theta$$

and that

$$\frac{1}{2\pi} \int_0^{2\pi} |f(re^{i\theta})|^2 \, d\theta = \sum_{n=0}^{\infty} |a_n|^2 \, r^{2n}.$$

(Note that $|w|^2 = w\bar{w}$.)

(iii) Deduce that $\sum_{n=0}^{\infty} |a_n|^2 r^{2n} \leqslant [M(r)]^2$.

(iv) Show that for all polynomials p of fixed degree m, the integral

$$\frac{1}{2\pi} \int_0^{2\pi} |f(e^{i\theta}) - p(e^{i\theta})|^2 \, d\theta$$

is a minimum when $p(z) = a_0 + a_1 z + a_2 z^2 + \cdots + a_m z^m$ and that this minimum value is $\sum_{k=m+1}^{\infty} |a_k|^2$.

Solutions

1. (i) We have $f^{(n)}(z) = \dfrac{(-1)^n n!}{(1+z)^{n+1}}$, and so $f^{(n)}(0) = (-1)^n n!$. Thus

$$\frac{1}{1+z} = \sum_{n=0}^{\infty} (-1)^n z^n.$$

The radius of convergence is 1, by the ratio test.

(ii) We have $f^{(n)}(z) = \dfrac{(-1)^n n!}{(1+z)^{n+1}}$, and so $f^{(n)}(1) = \dfrac{(-1)^n n!}{2^{n+1}}$. Thus

$$\frac{1}{1+z} = \sum_{n=0}^{\infty} \frac{(-1)^n (z-1)^n}{2^{n+1}}.$$

The radius of convergence is 2, by the ratio test.

(iii) We have $f^{(n)}(z) = \dfrac{(-1)^n n!}{z^{n+1}}$ and so $f^{(n)}(1) = (-1)^n n!$. Thus

$$\frac{1}{z} = \sum_{n=0}^{\infty} (-1)^n (z-1)^n.$$

The radius of convergence is 1, by the ratio test.

(iv) We have $f^{(n)}(z) = \dfrac{(n+1)!}{(1-z)^{n+2}}$, and so $f^{(n)}(0) = (n+1)!$. Thus

$$\frac{1}{(1-z)^2} = \sum_{n=0}^{\infty} (n+1) z^n.$$

The radius of convergence is 1, by the ratio test.

(v) We have $f^{(n)}(z) = \dfrac{(n+1)!}{(1-z)^{n+2}}$ and so $f^{(n)}(-1) = \dfrac{(n+1)!}{2^{n+2}}$. Thus

$$\frac{1}{(1-z)^2} = \sum_{n=0}^{\infty} \frac{n+1}{2^{n+2}}(z+1)^n.$$

The radius of convergence is 2, by the ratio test.

(vi) We have $f^{(n)}(z) = e^z$, and so $f^{(n)}(0) = 1$. Thus

$$e^z = \sum_{n=0}^{\infty} \frac{z^n}{n!}.$$

The series is convergent for all z, by the ratio test. (See Problem 1 of Section 6.4.)

(vii) The function $f(z) = \exp(k \operatorname{Log}(1+z))$ is the principal branch of $f(z) = (1+z)^k$ which has domain $\{z : z \in \mathbf{C}, z \neq \alpha$ where α is negative and less than or equal to $-1\}$(Fig. 6), and $f^{(n)}(z) = k(k-1)\ldots(k-(n-1))(1+z)^{k-n}$ for all integers n if k is not a positive integer.

So $f^{(n)}(0) = k(k-1)\ldots(k-n+1)$. Thus

$$f(z) = \sum_{n=0}^{\infty} \frac{k(k-1)\ldots(k-n+1)}{n!} z^n.$$

The radius of convergence is 1, by the ratio test.

If k is a positive integer, we get the familiar binomial expansion,

$$f(z) = \sum_{n=0}^{k} \binom{k}{n} z^n, \quad z \in \mathbf{C}.$$

In each case the radius of convergence is what we expect since if k is a positive integer, the function is entire; if k is not a positive integer things go badly wrong at -1.

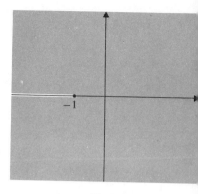

Fig. 6

2. If $f(z) = \displaystyle\sum_{n=0}^{\infty} a_n z^n$ is a solution, we must have

$$\sum_{n=2}^{\infty} n(n-1)a_n z^{n-2} + \sum_{n=0}^{\infty} a_n z^n = 0.$$

Thus $\displaystyle\sum_{n=0}^{\infty} b_n z^n = 0$, where

$$b_0 = 2 \cdot 1 \cdot a_2 + a_0$$
$$b_1 = 3 \cdot 2 \cdot a_3 + a_1$$
$$\cdots$$
$$b_n = (n+2)(n+1)a_{n+2} + a_n.$$

Thus $b_n = 0$ for all n, by the corollary to Theorem 9, because every Taylor coefficient of the zero function is equal to zero. Thus, if n is even, we have

$$a_{n+2} = -\frac{a_n}{(n+2)(n+1)}$$

$$= (-1)^2 \frac{a_{n-2}}{(n+2)(n+1)(n)(n-1)}$$

$$\cdots$$

$$= (-1)^{(n+2)/2} \frac{a_0}{(n+2)!}.$$

(We have expressed a_{n+2} in terms of a_0 after $(n+2)/2$ steps.) If n is odd, by similar reasoning, we have

$$a_{n+2} = (-1)^{(n+1)/2} \frac{a_1}{(n+2)!}.$$

Thus

$$f(z) = a_0 \sum_{n=0}^{\infty} \frac{(-1)^n}{(2n)!} z^{2n} + a_1 \sum_{n=0}^{\infty} \frac{(-1)^n}{(2n+1)!} z^{2n+1}.$$

Each of these two series converges for all z. Term by term differentiation shows them each to be a solution, and so justifies our assumption of the existence of such solutions. (If you have studied M201, Linear Mathematics, you will also see that they are linearly independent of each other and so $f(z)$ is the general solution of the equation.)

3. (i) $\displaystyle\sum_{k=0}^{n} a_k b_{n-k}$.

(ii) The functions f and g are analytic on D and so, therefore, is fg. We can therefore "expand" fg in a Taylor series:

$$(fg)(z) = \sum_{n=0}^{\infty} \frac{(fg)^{(n)}(0)}{n!} z^n.$$

(Notice how the notion of an analytic function again simplifies things: compare this proof of the convergence of the Cauchy product with that in real analysis.)

As in real analysis, the nth derivative of fg can be calculated by *Leibniz's formula*:

$$(fg)^{(n)}(z) = \sum_{k=0}^{n} \frac{n!}{k!(n-k)!} f^{(k)}(z) g^{(n-k)}(z).$$

Thus

$$(fg)^{(n)}(0) = \sum_{k=0}^{n} \frac{n!}{k!(n-k)!} f^{(k)}(0) g^{(n-k)}(0)$$

and

$$\frac{(fg)^{(n)}(0)}{n!} = \sum_{k=0}^{n} \frac{f^{(k)}(0)}{k!} \frac{g^{(n-k)}(0)}{(n-k)!}$$

$$= \sum_{k=0}^{n} a_k b_{n-k}$$

$$= c_n.$$

Thus $\displaystyle\sum_{n=0}^{\infty} c_n z^n$ is the Taylor series for fg in D.

(iii) From Problem 1, (iv) and (vi), and part (ii) of this problem,

$$\frac{e^z}{(1-z)^2} = \left(1 + z + \frac{z^2}{2!} + \frac{z^3}{3!} + \cdots\right) \cdot (1 + 2z + 3z^2 + 4z^3 + \cdots)$$

$$= 1 + 3z + \tfrac{11}{2}z^2 + \tfrac{49}{6}z^3 + \cdots.$$

Since the series for e^z and $\dfrac{1}{(1-z)^2}$ converge in the disc $|z| < 1$, so does this product series.

4. (i) $\displaystyle\sum_{k=0}^{n} a_k b_{n-k}$ is the coefficient of z^n in the Cauchy product of the Taylor series for fg. Since $f(z)g(z) = 1$, these coefficients are all zero, except the first, which is one.

(ii) Writing $g(z) = 1 + z$ in (i), we have $b_0 = b_1 = 1$, $b_n = 0$, $n > 1$. Thus $a_0 = 1$, $a_1 + a_0 = 0$, $a_2 + a_1 = 0$, etc., and so $a_0 = 1$, $a_1 = -1$, $a_2 = 1$, etc.

Thus, $\dfrac{1}{1+z} = 1 - z + z^2 - z^3 + \cdots$.

(iii) We have $\cos z = 1 - \dfrac{z^2}{2!} + \dfrac{z^4}{4!} - \cdots$.

Writing $f(z) = \cos z$ and $g(z) = \dfrac{1}{\cos z}$ in (i), we have

$$a_0 = 1, \, a_1 = 0, \, a_2 = -\tfrac{1}{2}, \text{ etc.}$$

Thus, $b_0 = 1$, $b_1 = 0$, $b_2 = \tfrac{1}{2}$, $b_3 = 0$, $b_4 = \tfrac{5}{24}$, etc., and so

$$\frac{1}{\cos z} = 1 + \tfrac{1}{2}z^2 + \tfrac{5}{24}z^4 + \cdots.$$

Hence

$$\tan z = \sin z \left(\frac{1}{\cos z}\right)$$

$$= \left(z - \frac{z^3}{3!} + \frac{z^5}{5!} + \cdots\right) \cdot (1 + \tfrac{1}{2}z^2 + \tfrac{5}{24}z^4 + \cdots)$$

$$= z + \tfrac{1}{3}z^3 + \tfrac{2}{15}z^5 + \cdots.$$

138

5. (i) We have $f^{(k)}(z) = \dfrac{k!}{(1-z)^{k+1}}$.

Thus

$$a_k = \frac{f^{(k)}(\alpha)}{k!} = \frac{1}{(1-\alpha)^{k+1}},$$

and so the Taylor series is as given.

(ii) By the ratio test, the series converges for $|z - \alpha| < |1 - \alpha|$, and so the radius of convergence is $|1 - \alpha|$.

(iii) The series $\dfrac{1}{1-\alpha} \sum\limits_{k=0}^{\infty} w^k$ is a geometric series, converging to

$$\frac{1}{1-\alpha} \cdot \frac{1}{1-w} = \frac{1}{1-z} = f(z).$$

(iv) By suitable choice of α, any point other than 1 can be included in the disc of convergence of the series and we know from results on power series that the sum function is analytic on every disc on which the series converges.

6. (i) We have

$$a_n = \frac{f^{(n)}(0)}{n!} = \frac{1}{2\pi i} \int_C \frac{f(z)}{z^{n+1}}\, dz,$$

by Cauchy's Formula for derivatives, where C is the circle $\{z : |z| = r\}$.

Thus, by the Estimation Theorem, $|a_n| \leqslant \dfrac{1}{2\pi} ML$, where M is the maximum of

$z \longrightarrow \left| \dfrac{f(z)}{z^{n+1}} \right|$ on C and L is the length of C. Since $M = \dfrac{M(r)}{r^{n+1}}$ and $L = 2\pi r$, we deduce that

$$|a_n| \leqslant \frac{M(r)}{r^n}.$$

(ii) Using the parametrisation $\gamma(\theta) = re^{i\theta}$, $\theta \in [0, 2\pi]$, we have

$$a_n = \frac{1}{2\pi i} \int_C \frac{f(z)}{z^{n+1}}\, dz = \frac{1}{2\pi i} \int_0^{2\pi} f(re^{i\theta}) r^{-(n+1)} e^{-(n+1)i\theta} rie^{i\theta}\, d\theta$$

$$= \frac{1}{2\pi r^n} \int_0^{2\pi} f(re^{i\theta}) e^{-ni\theta}\, d\theta, \quad \text{as required.}$$

Also

$$|f(re^{i\theta})|^2 = f(re^{i\theta}) \overline{f(re^{i\theta})}$$

$$= \sum_{n=0}^{\infty} a_n r^n e^{ni\theta} \sum_{k=0}^{\infty} \overline{a_k} r^k (e^{-ki\theta}).$$

Both series are absolutely convergent and so we can form the Cauchy product. The resulting series is the sum of a power series in $z = re^{i\theta}$ and $\bar{z} = re^{-i\theta}$. Both z and $\bar{z}$ are inside the circle of convergence and so we can integrate term by term. (One integral is with respect to z and one with respect to $\bar{z}$; or, effectively one with respect to θ and one with respect to $-\theta$.) The integrals will be of the form:

$$\int_0^{2\pi} e^{ni\theta} e^{-ki\theta}\, d\theta,$$

and are zero when $k \neq n$.

When $k = n$ we get 2π and so we deduce that

$$\int_0^{2\pi} |f(re^{i\theta})|^2\, d\theta = \sum_{n=0}^{\infty} r^{2n} a_n \overline{a_n} 2\pi$$

and so

$$\frac{1}{2\pi} \int_0^{2\pi} |f(re^{i\theta})|^2\, d\theta = \sum_{n=0}^{\infty} r^{2n} |a_n|^2, \quad \text{as required.}$$

139

(iii) Using the result from real analysis that

$$\int_a^b |f(x)|dx \leq |b - a|\sup\{|f(x)| : x \in [a, b]\},$$

we deduce that

$$\sum_{n=0}^{\infty} r^{2n}|a_n|^2 \leq [M(r)]^2.$$

(iv) Let $g = f - p$; then the integral in question is

$$\frac{1}{2\pi}\int_0^{2\pi} |g(e^{i\theta})|^2 \, d\theta = \sum_{n=0}^{\infty} |b_n|^2,$$

from (ii) with $r = 1$, where the b_n's are the Taylor coefficients of g. For $n > m$, $b_n = a_n$ because $g^{(n)} = f^{(n)}$ for $n > m$. Thus for $n > m$ the Taylor coefficients are not ours for the choosing (we can play only with m). But we can make $|b_n|^2 = 0$ for all $n \leq m$ by choosing

$$p(z) = a_0 + a_1 z + a_2 z^2 + \cdots + a_m z^m.$$

Hence, the minimum value of the integral is $\displaystyle\sum_{k=m+1}^{\infty} |a_k|^2$.

(Note: If you have studied MST282, Mechanics and Applied Calculus, or M201, Linear Mathematics, you may well have noticed the connection between Taylor series and Fourier series hinted at in this problem. If we write $z = re^{i\theta}$, then $\displaystyle\sum_{n=0}^{\infty} a_n z^n$ becomes $\displaystyle\sum_{n=0}^{\infty} a_n r^n e^{ni\theta} = \sum_{n=0}^{\infty} b_n e^{ni\theta}$, where $b_n = a_n k^n$, on the circle $\{z : |z| = k\}$. The problem of the convergence of Fourier series in an interval is thus one of convergence of Taylor series on a circle. It is when this circle happens to be the circle of convergence of the Taylor series that the discussion of Fourier series becomes difficult.)

6.7 UNIQUENESS

Towards the end of Section 6.5 we remarked on the extra understanding of the real Taylor series

$$\frac{1}{1 + x^2} = 1 - x^2 + x^4 - x^6 + \cdots, \quad |x| < 1,$$

that we obtain by considering the complex Taylor series

$$\frac{1}{1 + z^2} = 1 - z^2 + z^4 - z^6 + \cdots, \quad |z| < 1.$$

Effectively we are suggesting that a function defined by a real power series can be extended to a region in the complex plane by replacing x by z. It is clear that we should ask ourselves if such an extension is unique. This raises a broader question, which we may formulate thus:

Question

If two functions f and g are analytic on a region R and equal on a set S must they be equal everywhere in R?

As it stands the question is not well formulated since we must clearly say something about the set S. However it seems very likely that if S has a sufficient number of points then the answer to the question will be in the affirmative. We will now attempt to substantiate this guess.

$$* \qquad * \qquad * \qquad * \qquad * \qquad * \qquad * \qquad *$$

First we should notice that the question essentially asks on what kind of set an analytic function can be zero, for if $f = g$ on S then $h = g - f = 0$ on S. This is a very common step in discussing such problems. Let us therefore restrict our discussion initially to this aspect of the question.

Suppose then that h is analytic on a region R, and that S is a circle. We can see at once that if $h(z) = 0$ for $z \in S$, then, for all α inside S,

$$h(\alpha) = \frac{1}{2\pi i} \int_S \frac{h(z)}{z - \alpha} \, dz, \quad \text{by Cauchy's Formula,}$$

$$= 0, \quad \text{since } h(z) = 0 \text{ on } S.$$

So, if $h(z) = 0$ for $z \in S$, where S is a circle in R, then $h(\alpha) = 0$ inside S. But what about the remaining points of R?

Also we should consider sets other than the circle for S. Is it possible for $h(z)$ to be zero on some general simple-closed contour S in R and yet be non-zero at some other points of R? Is it possible for $h(z)$ to be zero on some general open set S in R and yet be non-zero at some other points of R? We introduce some new terminology which will assist our investigation.

Definition

If h is a function analytic on a region R and $h(\alpha) = 0$ for some $\alpha \in R$, then α is said to be a **zero** of h. The zero at α is said to be **of order** $n, n \geqslant 1$, if

$$h^{(k)}(\alpha) = 0, k = 0, 1, \cdots, n - 1, \text{ but } h^{(n)}(\alpha) \neq 0,$$

or, equivalently, if

$$h(z) = \sum_{k=0}^{\infty} a_k(z - \alpha)^k \text{ and } a_k = 0, k < n, \text{ but } a_n \neq 0.$$

A zero of order 1 is sometimes called a **simple zero**.

141

If $h^{(k)}(\alpha) = 0$ for $k = 0, 1, \ldots, n - 1$, then the first n coefficients of the Taylor series for h at α are zero and we can use Theorem 13 to deduce the following theorem:

Theorem 14

Let h be analytic on a region R, and let $\alpha \in R$ be a zero of order n of h, then

$$h(z) = (z - \alpha)^n g_n(z)$$

where g_n is analytic on R and $g_n(\alpha) \neq 0$.

This theorem is a good example of a global result: it tells us something about h on the whole of R.

We can now deduce quite easily that the zeros of finite order must be isolated, for if α is a zero of order n, then

$$h(z) = (z - \alpha)^n g_n(z)$$

and $g_n(\alpha) \neq 0$. But g_n is analytic and therefore continuous at α, and so $g_n(z)$ is non-zero on some neighbourhood of α. Thus $h(z)$ is non-zero on some neighbourhood of α except, of course, at α itself.

We now show that an analytic function, with the exception of the zero function, can have only zeros of finite order.

Theorem 15

If h is analytic on a region R and $h \neq 0$ on R then any zero of h on R is of finite order.

Proof

We prove this theorem by contradiction—showing that if for some $\alpha \in R$, $h^{(k)}(\alpha) = 0$, $k = 0, 1, 2, \ldots$, then $h(z) = 0$ for all $z \in R$. Now we know that h can be expanded as a Taylor series centred at α so that under these conditions $h(z) = 0$ for all z in some disc centre α. We have to show that $h(z)$ is also zero in the rest of R. We can argue as follows.

If β is any point in R then, since R is connected, we can join α and β by a polygonal path Γ. Suppose for the moment that we can choose Γ to be a line segment. In some disc D_0, centre α, $h = 0$.

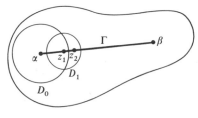

Fig. 7

Choose a point z_1 in $D_0 \cap \Gamma$. We know that $h^{(k)}(z_1) = 0$ for all positive integers k and so $h = 0$ on some disc D_1, centre z_1. Now assume we can choose z_2 in $D_1 \cap \Gamma$ but not in D_0 and repeat the argument. We now continue this argument and, by choosing the z_i's and D_i's appropriately, the point β will eventually lie in one of the discs and so $h(\beta) = 0$, which is a contradiction. If α and β cannot be joined by a line segment they can certainly be joined by a polygonal path and so the argument that follows can be continued successively along each segment of this path until we reach β.

Before we can give a method of choosing z_i and D_i we have to see just how large we can make the discs. We know from Taylor's Theorem that the discs can be made as large as we please as long as they are still contained in R. Let $d(z)$ be the distance from $z \in \Gamma$ to $\mathbf{C} - R$. Then from *Unit 2, Section 2.9* we know that d

142

is a continuous function and also that $d(z)$ has a minimum value. Call this minimum k. We know that wherever z_i is on Γ we can choose each disc D_i to have radius $k/2$, say. All that remains is to choose the centres. D_0 has centre α, radius $k/2$, and will intersect Γ at one point, ζ. Choose D_1 to have centre $\dfrac{\alpha + \zeta}{2}$ and radius $k/2$. D_1 will intersect Γ at ζ_1 say.

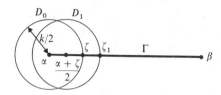

Fig. 8

Choose D_2 to have centre at $\dfrac{\zeta + \zeta_1}{2}$: D_2 and all subsequent discs (except the last) will intersect Γ in two points, but only one of these will not be contained in a previous disc. Suppose that for the disc D_i this point is ζ_i, then we choose D_{i+1} to have centre at $\dfrac{\zeta_{i-1} + \zeta_i}{2}$. (We leave it for you to show that there is a "last" disc.) ∎

Corollary 1

If h is analytic on a region R and $h \neq 0$ on R then the set of zeros of h has no cluster points in R (that is to say, any zero of h in R is isolated).

Proof

From Theorem 15 we deduce that, since h is not the zero function on R, any zero, α, of h is of finite order and so we can write

$$h(z) = (z - \alpha)^n g(z)$$

where g is analytic on R and $g(\alpha) \neq 0$. Since g is continuous we can find a neighbourhood of α on which $g(z)$, and hence $h(z)$, is non-zero. It follows that α is not a cluster point of the set of zeros, and so the zeros are isolated. ∎

Corollary 2

If h is analytic on a region R and $h = 0$ on a set S with a cluster point in R then $h = 0$ on R.

The second corollary, which follows immediately from Corollary 1, is the final answer to our question (page 141), for if $h = 0$ on a set with a cluster point in R then we can be sure that $h = 0$ on the whole of R. Thus if two functions f and g are analytic on R and $f = g$ on a set S with a cluster point in R, then $f = g$ on R. We state this result formally as Theorem 16.

Theorem 16 (The Uniqueness Theorem)

Let f and g be analytic on a region R, S be a subset of R with a cluster point in R and $f = g$ on S. Then $f = g$ on R.

Proof

Let $h = f - g$ and Z be the set of zeros of h in R, then $S \subset Z$. It follows from Corollary 2 that $h = 0$. ∎

As a simple corollary to the Theorem we have:

Corollary 1

If f and g are analytic on a region R and if $f = g$ on a line segment or a non-empty open set in R then $f = g$ on R.

143

Proof

Let $[z_1, z_2] \subset R$ and $f(z) = g(z)$ for $z \in [z_1, z_2]$. Then z_1 is a cluster point of the sequence $\{w_n\}$ where $w_n = z_1 + (z_2 - z_1)/n$.

If $f = g$ on an open set then they agree on an open disc in this set and therefore on a line segment in this open disc. ∎

Corollary 2

If f and g are analytic on a region R and $f^{(k)}(\alpha) = g^{(k)}(\alpha)$, $k = 0, 1, 2, \ldots$, for a point $\alpha \in R$ then $f(z) = g(z)$ for all $z \in R$.

Proof

The functions f and g will have the same Taylor series in an open disc centre α and so will agree on an open subset of R. The result follows from the previous corollary. ∎

Notice that in the proof of Corollary 1 to Theorem 15, α is an interior point of R. If a function is zero at an infinite number of points in a bounded region then we know from the Bolzano-Weierstrass Theorem that these points have a cluster point. We can now be sure that the function in question cannot be analytic at this cluster point. This fact can sometimes be used to deduce that certain functions fail to be analytic at various points.

Example

The function $z \longrightarrow \sin 1/z$ is not defined at $z = 0$, but the above result tells us that however we choose to define this function at 0, we cannot produce a function analytic on a neighbourhood of 0. This follows because $\sin 1/z = 0$ if $z = 1/(n\pi), n = \pm 1, \pm 2, \ldots$, and 0 is a cluster point of the set $S = \{z : z = 1/(n\pi)\}$.

Theorem 16 tells us that a function is uniquely determined on a region once we know its values on a convergent sequence with limit in the region. This may be compared with Cauchy's Formula which also tells us that a function is completely specified by its values on a subset of its domain. But whereas Cauchy's Formula gives a method of calculating the function at other points, the result of Theorem 16 is entirely theoretical as it stands.

An immediate practical pay-off from Theorem 16 is that if f is a function that reduces to a known real-valued function on the real axis for which we know the Taylor series, then the Taylor series (centred at a point on the real axis) of the complex function can be obtained particularly easily. Thus, because

$$\sin x = x - \frac{x^3}{3!} + \cdots, \quad x \in \mathbf{R},$$

we can deduce immediately that

$$\sin z = z - \frac{z^3}{3!} + \cdots, \quad z \in \mathbf{C}.$$

In the same way

$$\cos z = 1 - \frac{z^2}{2!} + \frac{z^4}{4!} - \cdots, \quad z \in \mathbf{C};$$

$$\exp z = 1 + \frac{z^2}{2!} + \frac{z^3}{3!} + \cdots, \quad z \in \mathbf{C}.$$

(Alternatively, these series give us a way of defining these complex functions.)

It is interesting and useful to note that not only do these series carry over to the complex plane but so also do some important relations such as $\sin^2 x + \cos^2 x = 1$. We can see this by defining the function F as $z \longrightarrow \sin^2 z + \cos^2 z - 1$. On the real axis $F = 0$ and since F is analytic on $\mathbf{C}$ we can deduce that $F = 0$ on $\mathbf{C}$.

This idea can be extended to formulas such as

$$\sin(x_1 + x_2) = \sin x_1 \cos x_2 + \cos x_1 \sin x_2$$

which involve two real variables. If we keep x_2 fixed, and define the function

$$G : z \longrightarrow \sin(z + x_2) - \sin z \cos x_2 - \cos z \sin x_2,$$

then G is analytic on $\mathbf{C}$ and zero on the real axis and so zero on $\mathbf{C}$. Now fix $z = z_1$ and define the function

$$H : z \longrightarrow \sin(z_1 + z) - \sin z_1 \cos z - \cos z_1 \sin z.$$

The function H is analytic on $\mathbf{C}$ and for z real we can choose $z = x_2$; we get

$$H(x_2) = G(z_1) = 0.$$

Thus (since x_2 can be any real number) H is zero on R and so zero on $\mathbf{C}$ (because z_1 can be any complex number) and so

$$\sin(z_1 + z_2) - \sin z_1 \cos z_2 - \cos z_1 \sin z_2 = 0$$

for all $z_1, z_2 \in \mathbf{C}$.

In a similar way, we can prove that

$$\exp(z_1 + z_2) = \exp z_1 \cdot \exp z_2,$$

and so on.

Summary

In this section we have proved that if $\{z_n\}$ is a sequence of points in a region R that converges to $\alpha \in R$, and f and g are analytic on R with $f(z_n) = g(z_n)$ for all n then $f(z) = g(z)$ for all z in R. As a corollary, if f and g agree on any line segment in R, or any non-empty open subset in R, then they agree throughout R. It follows that if $f^{(k)}(\alpha) = g^{(k)}(\alpha)$, $k = 0, 1, 2, \ldots$, for any given $\alpha \in R$, then f and g agree on the whole of R.

These results provide a method of defining complex functions that reduce to known real functions on the real axis and extending real Taylor series to complex Taylor series. We also saw (although we did not prove) a general result that functional relations (such as the standard trigonometric formulas) that are known for real functions hold also for the extended functions.

We also saw how to define the *order* of a zero of a function and that if a function is analytic on a region R and has a zero of infinite order at an interior point of R then the function must be the zero function on R.

Any zero of an analytic function is isolated, in the sense that if α is a zero then there is a punctured disc $\{z : 0 < |z - \alpha| < k\}$ on which $f(z)$ is non-zero (provided that f is not the zero function).

Self-Assessment Questions

1. Locate any zeros of the following functions and write down the order of each zero.

   (i) $f: z \longrightarrow \dfrac{z - \alpha}{z - \beta}$.

   (ii) $f: z \longrightarrow e^z$.

   (iii) $f: z \longrightarrow z^3(z - 3)$.

   (iv) fg where f has a zero of order h at α and g has a zero of order k at α.

   (v) $\dfrac{f}{g}$ where f and g are as in (iv) with $k < h$.

2. Write down two conditions each of which is sufficient to ensure that $f(z) = g(z)$ for all z in a region R on which both f and g are analytic.

3. Classify the following statements as True or False.

   (i) If $f(z) = (z - \alpha)^m g(z)$ and g is analytic on a region R, then f has a zero of order m at α.

   (ii) If f has a zero at α, then there is a function g such that $f(z) = (z - \alpha)^m g(z)$ for some integer m, and $g(\alpha) \neq 0$.

   (iii) If f is analytic on $\{z : |z| < 1\}$ and $f(1 - 1/n) = 0$ for all $n = 1, 2, 3, \ldots,$ then f is zero throughout $\{z : |z| < 1\}$.

Solutions

1. (i) α, order 1. (ii) No zeros. (iii) 0, order 3, and 3, order 1. (iv) α, order $h + k$. (v) f/g is not defined at α, and so does not have a zero there.

2. Each of the following statements is a sufficient condition.

   (a) $f(z_n) = g(z_n)$ for all z_n where $\{z_n\}$ is a sequence in R with limit in R.

   (b) $f^{(n)}(\alpha) = g^{(n)}(\alpha)$ for $n = 0, 1, 2, \ldots,$ at a point $\alpha \in R$,

   (c) $f = g$ on a non-empty open subset of R,

   (d) $f = g$ on a line segment in R.

3. (i) False: g could be the zero function.

   (ii) False: f must not be the zero function.

   (iii) False: although the sequence $\{1 - 1/n\}$ has a limit, this limit is not in the open disc.

6.8 PROBLEMS

1. Expand the following functions as power series centred at 0, *without* using the formula for the coefficients.

   (i) $f(z) = \dfrac{2}{(1 - z)^3}$;

   (ii) $f(z) = \text{Log}(1 + z)$;

   (iii) $f(z) = \displaystyle\int_0^z \phi(w)\, dw$, where $\phi(w) = \begin{cases} \dfrac{\sin w}{w}, & w \neq 0 \\ 1, & w = 0. \end{cases}$

2. (l'Hôpital's Rule)*. Let f and g be analytic on a region R, and let $f \neq 0$, $g \neq 0$. If $f(\alpha) = g(\alpha) = 0$ for some $\alpha \in R$, show that

$$\lim_{z \to \alpha} \frac{f(z)}{g(z)} = \lim_{z \to \alpha} \frac{f'(z)}{g'(z)}$$

   when these limits exist, and if one limit does not exist, then neither does the other.

3. Let f be analytic on a region R containing the point 0 and let $f(1/n)$ be real and $f(1/n) < 1/2^n$ for $n = 1, 2, 3, \ldots$. Prove that $f = 0$ on R.

4. (i) Let f be analytic on $\mathbf{C}$ and real-valued on the interval $(-\varepsilon, \varepsilon)$ of the real axis. Prove that f is real-valued on the whole real axis and that $f(\bar{z}) = \overline{f(z)}$ for all $z \in \mathbf{C}$.

   (ii) Deduce that if f is any one of exp, sin, cos, sinh or cosh then $f(\bar{z}) = \overline{f(z)}$. (See *Unit 3*, pages 122 and 127.)

5. Let f be a function which is real on the real axis and imaginary on the imaginary axis. Show that f is an odd function. Find a similar condition which guarantees that f is an even function.

6. Show that the function $f(z) = \frac{1}{2}[\exp(iz) + \exp(-iz)]$ is the unique extension of the real function cos to the complex plane.

7. Find the orders of the zeros at $z = 0$ of
   (i) $z \longrightarrow z(e^2 - 1)$,

   (ii) $z \longrightarrow z^2(\cos z - 1)$,

   (iii) $z \longrightarrow 6 \sin(z^2) + z^2(z^4 - 6)$.

8. Is there a function f analytic on a region containing 0 such that $f(x) = \exp(-1/x^2)$ when x is real and not zero?

9. Give an example of a function f analytic on a bounded region R which has infinitely many zeros in R but is not identically zero on R.

10. The function f given by

$$z \longrightarrow \begin{cases} z \sin \dfrac{1}{z}, & z \neq 0 \\ 0, & z = 0, \end{cases}$$

   is continuous at 0. Is there any region containing 0, on which the function is analytic?

Guillaume l'Hôpital (1661–1704) spent most of his life in Paris and learnt calculus from John Bernoulli, who was one of the only four people at the time who understood it (the others being his brother James, Leibniz and Newton)! He wrote the first ever text-book on calculus, and also published a book on conics, which became a standard work on the subject for nearly a century.

Solutions

1. (i) From real analysis

$$\frac{1}{1-x} = 1 + x + x^2 + \cdots, \quad |x| < 1,$$

and so in the complex plane

$$\frac{1}{1-z} = 1 + z + z^2 + \cdots, \quad |z| < 1,$$

because the function $z \longrightarrow 1/(1-z)$ is analytic on the disc $|z| < 1$. Differentiating, we get

$$\frac{1}{(1-z)^2} = 1 + 2z + 3z^2 + \cdots, \quad |z| < 1,$$

and

$$\frac{2}{(1-z)^3} = 2 + 6z + 12z^2 + \cdots, \quad |z| < 1.$$

(ii) From real analysis

$$\frac{1}{1+x} = 1 - x + x^2 - \cdots, \quad |x| < 1,$$

and so in the complex plane

$$\frac{1}{1+z} = 1 - z + z^2 - \cdots, \quad |z| < 1,$$

because the function $z \longrightarrow 1/(1+z)$ is analytic on the disc $|z| < 1$. Integrating, we get

$$\mathrm{Log}(1+z) = z - \frac{z^2}{2} + \frac{z^3}{3} - \cdots, \quad |z| < 1,$$

since $\mathrm{Log}(1+z) = 0$ at $z = 0$.

(iii) We have

$$\frac{\sin w}{w} = 1 - \frac{w^2}{3!} + \frac{w^4}{5!} - \cdots \text{ for all } w \neq 0.$$

Therefore,

$$\int_0^z \phi(w)\,dw = z - \frac{z^3}{3\cdot 3!} + \frac{z^5}{5\cdot 5!} - \cdots, \quad \text{for all } z.$$

2. By Theorem 14, there are positive integers m and n and functions F and G analytic on R such that

$$f(z) = (z - \alpha)^m F(z) \quad \text{and} \quad g(z) = (z - \alpha)^n G(z),$$

where $F(\alpha) \neq 0$ and $G(\alpha) \neq 0$. For $z \neq \alpha$,

$$\frac{f(z)}{g(z)} = \frac{(z - \alpha)^m F(z)}{(z - \alpha)^n G(z)},$$

and so

$$\frac{f'(z)}{g'(z)} = \frac{(z - \alpha)^{m-1}[mF(z) + (z - \alpha)F'(z)]}{(z - \alpha)^{n-1}[nG(z) + (z - \alpha)G'(z)]}$$

$$= (z - \alpha)^{m-n}\left(\frac{mF(z) + (z - \alpha)F'(z)}{nG(z) + (z - \alpha)G'(z)}\right).$$

If $m = n$, then both limits are $\dfrac{F(\alpha)}{G(\alpha)}$.

If $m > n$, then both limits are zero.

If $m < n$, then the limits do not exist.

3. Since f is continuous it is clear that $f(0) = 0$. Thus by Theorem 14 we know that either $f = 0$ on R or there is a positive integer m and an analytic function g such that $g(0) \neq 0$ and

$$f(z) = z^m g(z).$$

Suppose this latter situation holds. Then

$$f(1/n) = \frac{1}{n^m} g(1/n) < \frac{1}{2^n}$$

and so

$$g(1/n) < \frac{n^m}{2^n}.$$

Since g is analytic and $\lim\limits_{n\to\infty} \dfrac{n^m}{2^n} = 0$, we have $g(0) = 0$. We thus obtain a contradiction and deduce that $f = 0$ on R.

4. (i) On $(-\varepsilon, \varepsilon)$ we have $f(x) = \sum\limits_{n=0}^{\infty} a_n x^n$ and from the uniqueness properties we know that the a_n's are the Taylor coefficients of f regarded as a real function and so the a_n's are real. Since the Taylor series for the complex function f converges throughout $\mathbf{C}$, we have $f(z) = \sum\limits_{n=0}^{\infty} a_n z^n$, $z \in \mathbf{C}$, and so $f(z)$ is real if z is real. Let z be a fixed point in $\mathbf{C}$, and define f_N by

$$f_N(z) = \sum_{n=0}^{N} a_n z^n.$$

Then, we know from the properties of polynomials that $f_N(\bar{z}) = \overline{f_N(z)}$ for all N. Thus, if $\operatorname{Re} f_N(z) = u_N$ and $\operatorname{Im} f_N(z) = v_N$, that is

$$f_N(z) = u_N + iv_N,$$

then

$$f_N(\bar{z}) = u_N - iv_N.$$

If $\lim\limits_{N\to\infty} u_N = u$ and $\lim\limits_{N\to\infty} v_N = v$, then

$$f(z) = u + iv$$

and

$$f(\bar{z}) = \lim_{N\to\infty} u_N - i \lim_{N\to\infty} v_N$$
$$= u - iv$$
$$= \overline{f(z)}.$$

(ii) Each of the functions is entire, that is analytic on $\mathbf{C}$, and real-valued on the real axis.

5. From Problem 4, since $f(z)$ is real for z real, the Taylor coefficients a_n of f are real. We have

$$f(iy) = \sum_{n=0}^{\infty} a_n (iy)^n$$
$$= \sum_{n=0}^{\infty} (-1)^n a_{2n} y^{2n} + i \sum_{n=0}^{\infty} (-1)^n a_{2n+1} y^{2n+1}. \qquad (*)$$

For $f(iy)$ to be imaginary, we must have $a_{2n} = 0$ for all n (an application of "uniqueness"). Thus

$$f(z) = \sum_{n=0}^{\infty} a_{2n+1} z^{2n+1}$$

and so $f(-z) = -f(z)$, that is, f is odd.

Similarly, if $f(z)$ is real for z real and for z imaginary, then from $(*)$ we must have $a_{2n+1} = 0$. In this case f is even.

6. We know from *Unit 3, Differentiation*, that the two functions agree on the real axis. If g were another complex function that agreed with the real function cos on the real axis (another extension of cos) then we would have $f = g$ on the real axis and so, by the uniqueness properties, $f = g$ everywhere.

7. (i) $z(e^z - 1) = z\left(z + \dfrac{z^2}{2} + \cdots\right)$

$$= z^2 \times \text{(an analytic function which is not zero at 0)}.$$

The zero at 0 is of order 2.

149

(ii) $z^2(\cos z - 1) = z^2\left(\dfrac{z^2}{2!} - \dfrac{z^4}{4!} + \cdots\right)$

$= z^4 \times$ (an analytic function which is not zero at 0).

The zero at 0 is of order 4.

(iii) $6\sin(z^2) + z^2(z^4 - 6) = 6\left(z^2 - \dfrac{z^6}{3!} + \dfrac{z^{10}}{5!} - \cdots\right) + z^6 - 6z^2$

$= z^{10} \times$ (an analytic function which is not zero at 0).

The zero at 0 is of order 10.

8. If such a function exists, it must have a Taylor series in some disc $|z| < k$ and this must correspond to the real Taylor series for the function

$$f(x) = \begin{cases} \exp(-1/x^2), & x \in (-k, 0) \cup (0, k) \\ 0, & x = 0. \end{cases}$$

But $f^{(r)}(0) = 0$ for all $r > 1$ and so each of the Taylor coefficients is zero.

So the Taylor series for f at 0 converges for all z, but its sum is $f(z)$ only at $z = 0$. Thus f cannot be extended to an analytic complex function in a region containing 0.

Some explanation for this strange behaviour is to be found by considering the behaviour of $\exp(-1/z^2)$ when z is imaginary. If $z = iy$, then $\exp(-1/z^2) = \exp(1/y^2)$. This becomes arbitrarily large as y gets small, contrasting with $\exp(-1/x^2)$ which becomes small as x gets small. The function $f(z) = \exp(-1/z^2)$ is clearly not continuous at the origin. This explains not only why the function cannot be extended analytically in a region containing 0 but is another illustration of how complex analysis can throw interesting light on some peculiar phenomena of real analysis (in this case the fact that the Taylor series for $f(x) = \exp(-1/x^2)$ at 0 converges to $f(x)$ only at 0).

9. $f : z \longrightarrow z \sin 1/z; \quad R = \{z : 0 < |z| < 1\}.$
    Notice that although $f = 0$ on a convergent sequence of points in R, we cannot conclude that $f = 0$ everywhere, because the limit of this sequence is not in R.

10. No, because $f(z) = 0$ at a sequence of points converging to a limit in the region. If f were analytic we could deduce that $f = 0$ everywhere in the region, and this is not the case.

    Similar remarks apply to the function obtained by replacing $z \sin 1/z$ by $z^k \sin 1/z$ where k is any positive integer. Recall that we try to make $x \longrightarrow x^k \sin 1/x$ better behaved at 0 (in the sense that higher order derivatives exist there) by increasing k. (See **Spivak**, Chapter 10). But whatever value one fixes for k, the $(k - 1)$th derivative will not exist. We know from complex analysis that if a function is analytic, then it is infinitely differentiable.

GLOSSARY OF SYMBOLS

The following items of notation are explained on the pages given. Most other items of notation used in the text may be found in one of the following: *Mathematical Handbook* for M100, The Mathematics Foundation Course; *Handbook* for M231, Analysis; *Units 1, 2 and 3* of M332, Complex Analysis.

$\int_a^b \phi, \int_a^b \phi(t)\, dt$ 13

$f(t)|_a^b$ 13

$\tilde{\gamma}$ 23

$\int_\gamma f, \int_\gamma f(z)\, dz$ 30

$\text{Wnd}\,(\gamma, \alpha)$ 35

$\int_\Gamma f, \int_\Gamma f(z)\, dz$ 38

$\int_{[a,b]} f$ 40

$\Gamma = (\Gamma_0, \ldots, \Gamma_{n-1})$ 41

$\Gamma + \Delta$ 43

$\Gamma = \Gamma_1 + \cdots + \Gamma_n$ 43

$-\Gamma$ 43

$\int_\Gamma |f(z)| \cdot |dz|$ 68

$\left(\int_C - \int_{C'} \right) f$ 76

$\lim_{z \to \infty} f(z) = l$ 90

$\int f, \int f(z)\, dz$ 96

$\int f(z)\, dz = F(z)$ 96

$\int_\alpha^\beta f(z)\, dz$ 96

$\{a_n\}$ 106, 107

$\lim_{n \to \infty} a_n = l$ 106, 107

INDEX

COMPLEX ANALYSIS

1 Complex Numbers

2 Continuous Functions

3 Differentiation

4 Integration

5 Cauchy's Theorem I

6 Taylor Series

7 NO TEXT

8 Singularities

9 Cauchy's Theorem II

10 The Calculus of Residues

11 Analytic Functions

12 NO TEXT

13 Properties of Analytic Functions

14 Laplace Transforms

15 Number Theory

16 Boundary Value Problems

Course Team

Chairman:	Dr. G. A. Read	Senior Lecturer
Authors:	Dr. P. D. Bacsich	Lecturer
	Dr. M. Crampin	Lecturer
	Mr. N. W. Gowar	Senior Lecturer
	Dr. R. J. Wilson	Lecturer
Editor:	Mr. R. J. Knight	Lecturer
B.B.C.:	Mr. D. Saunders	

With assistance from:

Mr. R. Clamp	B.B.C.
Mr. D. Goldrei	Course Assistant
Mr. H. Hoggan	B.B.C.
Mr. R. J. Margolis	Staff Tutor
Dr. A. R. Meetham	Staff Tutor
Mr. J. E. Phythian	Staff Tutor
Mrs. P. M. Shepheard Rogers	Course Assistant
Dr. C. A. Rowley	Course Assistant